Health and Difference

Studies of the Biosocial Society

General Editor: **Catherine Panter-Brick**, Professor of Anthropology, University of Durham

The Biosocial Society is an international academic society engaged in fostering understanding of human biological and social diversity. It draws its membership from a wide range of academic disciplines, particularly those engaged in 'boundary disciplines' at the intersection between the natural and social sciences, such as biocultural anthropology, medical sociology, demography, social medicine, the history of science and bioethics. The aim of this series is to promote interdisciplinary research on how biology and society interact to shape human experience and to serve as advanced texts for undergraduate and postgraduate students.

Volume 1
Race, Ethnicity and Nation: Perspectives from Kinship and Genetics
Edited by Peter Wade

Volume 2
Health, Risk and Adversity
Edited by Catherine Panter-Brick and Agustín Fuentes

Volume 3
Substitute Parents: Biological and Social Perspectives on Alloparenting in Human Societies
Edited by Gillian Bentley and Ruth Mace

Volume 4
Centralizing Fieldwork: Critical Perspectives from Primatology, Biological and Social Anthropology
Edited by Jeremy MacClancy and Agustín Fuentes

Volume 5
Human Diet and Nutrition in Biocultural Perspective: Past Meets Present
Edited by Tina Moffat and Tracy Prowse

Volume 6
Identity Politics and the New Genetics: Re/Creating Categories of Difference and Belonging
Edited by Katharina Schramm, David Skinner and Richard Rottenburg

Volume 7
Ethics in the Field: Contemporary Challenges
Edited by Jeremy MacClancy and Augstin Fuentes

Volume 8
Health and Difference: Rendering Human Variation in Colonial Engagements
Edited by Alexandra Widmer and Veronika Lipphardt

Health and Difference

Rendering Human Variation in Colonial Engagements

Edited by
Alexandra Widmer and Veronika Lipphardt

berghahn
NEW YORK • OXFORD
www.berghahnbooks.com

First published in 2016 by

Berghahn Books
www.berghahnbooks.com

Library of Congress Cataloging-in-Publication Data

A C.I.P. cataloging record is available from the Library of Congress

British Library Cataloguing in Publication Data

A catalogue record for this book is available from the British Library

Printed in the United States on acid-free paper.

ISBN 978-1-78533-271-5 (hardback)
ISBN 978-1-78533-272-2 (ebook)

Contents

List of Figures

Acknowledgements

We have been incredibly fortunate that so many people participated in bringing this volume to completion.

The volume has its origins in the June 2012 Workshop 'Colonial Subjects of Health and Difference: Races, Populations Diversities'. We would like to thank those present for their intellectual contributions that made it such a congenial and memorable event, in particular: Carlos Lopez Beltran, Daniel Bendix, Omnia El Shakry, Judith Littleton, Jin-Kyung Park, Julie Park, Joanna Radin, J. Emmanuel Raymundo, Katharina Schramm, Fenneke Sysling, Daniel Walther, Christine Winter and Ricardo Ventura Santos.

To those who contributed to the volume, thank you for all your intellectual labour and timely responses; we are honoured.

We are grateful to have had administrative and editorial assistance from Birgitta von Mallinckrodt, Ricky Heinitz, Luci Luft, Nina Ludwig, Leon Kokkoliadis, Eric Llavera and Sarah Fruendt.

The Max Planck Institute for the History of Science provided generous support for the workshop and the development of the volume.

Our collaboration began in the Max Planck Research group 'Twentieth Century Histories of Knowledge About Human Variation', led by Veronika Lipphardt. We would like to thank in particular Jenny Bangham, Susanne Bauer, Sarah Blacker, Staffan Mueller-Wille, Martina Schluender, Judith Kaplan, Yuriditzi Pasacio Montijo, Lara Keuck, Samuel Coghe, and all the guests who came for the many warm and insightful conversations that began during this time and have since continued.

The anonymous reviewers gave exceptionally helpful feedback. We wish to thank Catherine Panter-Brick, editor of the Biosocial Society Series at Berghahn Books, who has been incredibly supportive throughout. Working with Berghahn Books has been a delight.

To our families, who make our intellectual work possible and give our lives the richness that they do, our gratitude.

Alexandra Widmer and Veronika Lipphardt
November 2015

INTRODUCTION
Health and Difference:
Rendering Human Variation in Colonial Engagements

Veronika Lipphardt and Alexandra Widmer

Human variation, with its many dimensions, represented a fascinating topic of research for life scientists and posed challenging administrative issues for colonial bureaucrats in the first half of the twentieth century. For officials, managing the challenges of colonial administration was often dependent on acquiring data on their subject populations, while, conversely, the scientific pursuit of that data was firmly embedded in colonial rule. For people who were or became colonial subjects during this time, colonial rule meant, at the very least, struggling with new kinds of illnesses, expertise and exploitation. It also often meant being counted and categorized in the name of welfare and reform.

In this volume, contributors follow physicians, demographers, nutrition experts, physical anthropologists, colonial agents, military officials and missionaries in colonies all over the globe, drawing specific attention to how they tried to sort out pressing health problems of populations they perceived to be diverse. From these eight empirical contexts concerned with health and difference in colonial projects, we start from what others have shown, namely that scientific racism, racial medicine and colonial rule were linked in predictable but also ambiguous and contradictory ways (Ernst 1999; Tilley 2014).

Recently, scholars have recognized that a focus on race science or racial research is too narrow to understand how biological differences were made productive in colonial contexts. Helen Tilley (2014) has pointed to the fact that for colonial administrators, racial categories and racial hierarchies were often not the kind of information they could comfortably draw on and that racial research was not the most favoured scientific domain in the colonies. However, while Tilley correctly interprets this as the beginning of political maturity by a lessening adherence to racial ideology, another aspect is that actors developed a new attentiveness towards scientific inconsistencies or incongruities in local constellations. In the realities they encountered, differences did not necessarily map onto racial classifications, but instead seemed to point to other, more distinctive, more meaningful, or more clear-cut fixed differences. Hence, colonial administrators and scientists might have simply reacted to the one 'signal of resistance' (Fleck, Trenn and Merton 1979) that seemed strongest while documenting the situation. This includes the possibility that they set aside racial research, racial thinking and racial classification in the light of more convincing evidence in one particular instance, but maybe did not dismiss 'race' as a convincing explanation, classificatory tool or potential topic in other respects.

With a broad perspective, this volume explores various scientific attempts to locate order in or generate explanations for the overwhelmingly complex picture of human variation. Scientists did so by employing a multitude of categories and classifications from heterogeneous sources. As stated above, not all were meant to classify *racial* variation; some were centrally oriented towards assessing the variation of other biologically relevant aspects of population health such as the impact of different cultural lifestyles on nutrition and digestion. Attempts to standardize these locally dependent categories proved difficult. To be sure, racial categories were often more or less visibly implicated in these attempts. But still, it is noteworthy that not every life scientist who researched human variation – or who drew on human variation for researching other topics – was necessarily engaged in race science. Those scientists often employed multiple social and biological variables and categories of identity in their efforts to understand and improve the population socially, bodily and hygienically. 'Race' was only one aspect in the categorization of difference, and even where 'race' was an important consideration, it was often considered in relation to – or alongside – other biological and non-biological factors.

Accordingly, and as a contribution to the history of science, this collection builds on historians' previous work on studies of human difference in colonial contexts (e.g. Rich 1990; Ernst and Harris 1999; Dubow 2006; Douglas and Ballard 2008) in that it suggests a shift in focus by looking beyond the realm that is usually described as 'race science'. Many historical accounts of race science or scientific racism have been framed as histories of scientists' explicit interest in 'race', and have focused on how researchers generated 'racial' classifications or typologies measured through anthropometric or serological methods (Stepan 1982; Barkan 1992; Weindling 1993). The chapters here show how 'race' and racial categories surfaced as significant components of colonial research on other topics pertaining to the health of human groups, such as nutrition, demography, medicine and human biology. The chapters also show how all of these research fields employed 'race' in relation to other categories of human difference. In this way, the chapters show how 'race' stood for a firm belief in innate differences as rather fixed and unchangeable in relation to other kinds of more mutable differences. Addressing multiple categories of difference, we aim to embed the history of racial categories in a somewhat more broadly construed history of categorizing differences. At the same time, we acknowledge that notions of 'race' and racial categories played a crucial, if not structuring, role in any categorization of difference.

We highlight these shifts in focus in the history of science by analysing how such research was connected to colonial administrative requirements. Colonial administrators and researchers shared the perceived requirement of delineating a population and the differences within that population. The categories deemed administratively and socially relevant for dividing people in colonial contexts were entangled with the scientific categories and conceptual tools used in sciences of human diversity. Actors in both spheres shared common interests, resources and networks, practises and discourses. Following multiple categories of difference to a broad range of health concerns reveals that the sphere of colonial administration and the sphere of colonial medicine were indeed closely tied, adding to the significance of the categories of 'race' and 'culture' in colonial medicine that others have so clearly shown (e.g. Vaughan 1991; Anderson 2006).

The scientists' desire and capacity to document was matched by colonial officials' prolific output of reports, letters and other publications. Documentation was a central preoccupation in the bureaucratic routines of health administration (Widmer 2008). The countless reports of medical doctors (Bado, Coghe, Ehlers), military researchers (Mitchell), UN nutritionists (Galluzzi Bizzo), anthropological researchers (Cândido da Silva, Cooper, Pols) as well as correspondence from settlers concerned with social and cultural reform (what they might have called 'civilization'), and letters from various humanitarian groups in the colonies as well as in metropolitan centres piled high on the work tables of colonial administrators, as they pile high in archives today.

Colonial health administrations and medical researchers in their employ presented their statistical data in a wide variety of publications and reports. They assembled tables, graphs or maps to convey accuracy and in-depth coverage of minute details on the health status of populations and individuals subject to colonial rule. Sometimes these tables underlined the author's claim that the health situation of those populations was under control; sometimes they helped to express worries that control might get lost, or that it might not be possible to even establish control. In any case, knowing the numbers was one of the first steps for both researchers and administrators attempting to solve logistical or epistemic problems.

What these colonial tables or graphs represent, read in concert with related paper trails, are countless encounters and multiple and arduous forms of labour leading up to their publication on paper. Often enough bureaucrats or medical personnel failed to accomplish their intentions for medical care in the face of messy realities and few resources; but reports had to be written and results had to be presented. We want to emphasize that the classifications in these documents are not to be taken as a starting point from which to write a history. On the contrary, the categories are often ugly placeholders and poor witnesses of an extremely rich and troubled history of relationships between human beings. The classifications can be taken as a point to begin inquiries that point to the encounters and relationships that preceded the tables and census forms. Understood as dynamic and temporary placeholders, these classifications can help us to explore the many dimensions of the contingent and multiple kinds of practical historical relationships of colonialism. As Pels writes, such scientific classifications 'both indicate and obscure practical historical relationships.... The savage and the civilised never existed in separation: their definition itself already marked a relationship' (2008: 280). The discussions of classifications in this volume reveal some of those hitherto neglected practical relationships. Like the anthropological categories that Pels is writing of, the categories of human difference in the life sciences 'have to be understood as historical relationships before they can be treated as objects, tools or rules of the discipline' (ibid.).

Once these classifications and categories have landed on paper, it is well worth investigating their material, technical and processual lives. How were they part of the reorganization of social and political relations? From regulating labour standards, to the fight against sleeping sickness, what did the documents do, what did their material presence, their production, their circulation, their storage, their retrieval systems produce? The power and effect of colonial discourse is of course important, but given the imperative of both scientists and administrators for innovating styles and tools of documentation, the making and the circulation of documents is also a crucial activity. What did the creation of what colonial historian Sean Hawkins calls 'the world on paper' (2002) produce as it pertained to health administration? The documentary practices – makeshift though they might have been at times – were central to colonial statecraft and colonial forms of governmentality (though not exclusive to other ways of controlling territories) (Appadurai 1993; Cohn 1996; Prakash 1999; Dirks 2004). Together, the chapters show how the peculiarities of colonial bureaucratic routines, documentary practices and forms of governmentality were entangled with the questions and practices of life scientists. While the connection between anthropology, medicine and colonial governance has been made before, one of the contributions this volume makes is to show the breadth of overlapping bureaucratic and scientific practices aimed at categorizing and understanding human difference.

The knowledge, documentary and archival practices associated with health and medicine were but one – albeit very significant – dimension of the expansive range of colonial documentary and archival practices (e. g. Bayly 1996; Stoler 2009; Roque and Wagner 2012). The chapters thus provide insights into the complex relationship of colonial knowledge to power and governance writ large. Implicitly, then, the volume engages with the colonial dimensions of biopolitics, bureaucracy and governmentality. Foucault's well known work connecting population knowledge, power and governmentality animates this volume, but, as Stoler (1995) and Scott (1995) have pointed out long ago, his analysis did not extend to colonial contexts. With the volume's focus on the connection between colonial knowledge and health administration, it shows the extent to which colonial governmentality was accomplished through the documentary practices of categorizing difference. Furthermore, focusing on categorizing practices in association with discourses and scientific debates points to the fact that colonial knowledge and administration were profoundly material, shaping relationships between, for example, human flesh, swamps and mosquitos and how parents could feed their children.

As indicated above, another way that this collection adds to literatures on colonial medicine and the science of human variation is by making a shift in focus regarding the methodological approach: the chapters collected here concentrate on documentary practices, in relation to discourses, scientific debates or ideologies.

Through examining paper trails of bureaucratic procedures, artefacts of institutional life, alongside scientific practices, we can partially reconstruct how such practices rendered human difference in multiple ways. This volume looks at the effects of these documentary practices, not just in their content, but as a practice of documenting the world that resulted in figuring out population categories for marriage certificates, travel documents, census forms, demographic tables, statistics pertaining to hygiene or nutritional requirements, or plotting different populations on maps. Those population categories were more than just convenient names or labels: they were productive tools of marking difference, endowed with the power to impact the lives of many.

Scholars have told remarkable accounts of the history of knowledge in the topics discussed here. However, in following Bernard Cohn's approach (1996), we maintain that analysing documentary practices helps to reveal and understand 'more complex constellations of difference categories' (Ernst 1999: 8).[1] To be sure, how notions of race intersected with those of gender, class and nation has been an issue in many studies recently, and these difference categories are also crucially important in the chapters assembled here. But, informed by work on classifying, sorting and categorizing from Science and Technology Studies (Douglas and Hull 1992; Bowker and Star 2000), we deem it necessary to go beyond these ever important difference categories and to look also at processes of categorization that could be more fleeting or temporary. Taking Boris and Janssens' invitation to 'complicate categories' (1999) by paying attention to a more local and contingent level, documentary practices bring to the fore classifications and groupings that were first and foremost context dependent and local, and that only made sense in a very specific setting. For example, in Letícia Galluzzi Bizzo's chapter, the 'rice eaters' formed an actors' category in mid-twentieth-century nutritional science. International experts tried to map 'rice eaters' onto racial categories. They failed in these attempts and the categories paved the way for other divisions of people, other policies, and other epistemic objects than racial classifications would have done. 'Rice eaters' described not only a community with a diet mainly based on rice, but a category that figured in the practices of the FAO of making recommendations and standards for daily calorie requirements. It is through the practice of documentation that these categories, which originated in specific contexts, could aspire to relevance beyond one particular setting.

The contributors to this volume explore the troubling presence of actors' concepts of 'race' in a range of different kinds of colonial research on human variation between 1900 and 1950. They focus not just on notions of human difference in racial science, but also consider the administrative procedures and data collection practices that produced human difference in a wider constellation of epistemic questions about human variation. This volume thus opens new questions about how the contingencies of colonial documentary practices of health administration and those in the life sciences shaped knowledge of human variation. The chapters contribute to an understanding of medicine in specific colonial empires, as well as to the history of anthropology, demography and the life sciences. Taken together, they open new intellectual spaces on the range of colonial concerns regarding the administration of health and knowledge practices preoccupied with categorizing human difference. And yet, by looking at 'race' within these broader institutional and scientific frameworks, we do not intend to deny the harmful effects of race science. Rather, we want to draw specific attention to the research designs, methods and results that reveal the pervasive nature of racial thinking in research on human variation, and the colonial interventions with which it was entangled.

Social, cultural and ecological contexts at local and regional scales shaped the connections between the study of racial and cultural difference and the colonial administration of medicine. This has been well documented and analysed in such diverse places as in the Pacific Islands (Perez Hattori 2004; Anderson 2006, 2009), Africa (Tilley 2011) and India (Arnold 1993; Harrison 1999; Pande 2010; Deb Roy 2015). Related to such analyses, historians have also studied the entanglement of anthropology and the life sciences in European empires such as the British (e.g. Vaughan 1991), French (e.g. Osborne 2005), Portuguese (e.g. Roque 2010), Dutch (e.g. Stoler 1995), German empires (e.g. Zimmerman 2001). Indeed, because of the connections between colonial infrastructures, networks and ideologies, histories that focus on particular colonies or particular empires are methodologically practical and intellectually rigorous. In this volume, however, we present examples from across many empires and world regions and do not aim for a complete coverage of all relevant world regions. We see the breadth and scope of the chapters in terms of the empires – British, German, Dutch, French, Portuguese – and as well as the range of places – Nauru and the New Hebrides in the South Pacific, Angola, Brazil, Afrique occidentale française (AOF), British Uganda, Dutch Indies – as an opportunity. With regard, of course, to local specificities, this volume aims to respond to Warwick Anderson's provocation 'to look for what is colonial about Western medicine in any setting' (1998: 522). Furthermore, as Anderson makes clear, 'we need to recognize that the basic language of Western medicine, with its claims to universalism and modernity, has always used, as it still does, the vocabulary of empire' (ibid.: 529).

The time period covered by this volume starts in the late nineteenth century, focuses on the first half of the twentieth century and, in a few chapters, moves into the 1950s and early 1960s. It marks the heyday of colonial rule, but also a significant period in the history of medicine and the life sciences. Scientists in these fields conceived of human variation as a predominantly biological phenomenon, or they attempted to separate the biological from the cultural dimension to focus their investigations on the former. Hence, new bacteriological and biochemical methodologies were introduced in the study of human diseases and human variation. Even though the new methods would only become effective controls of disease in clinical contexts and public health much later (Quirke and Gaudillière 2008), in the beginning of the twentieth century the hopes of many rested on these new methodologies. Physiologists, biochemists and microbiologists produced new kinds of data – metabolic, transpiration or excretion rates for example – that all appeared to allow for studies of population difference. This growing stream of data added to classical data of human variation research, mainly achieved through anthropometrics, although seroanthropology (with blood group studies) and the new genetic thinking effectively challenged this methodology from the 1920s onwards (Lipphardt 2012, 2014). In all those fields, the 1950s brought decisive changes. Research on human variation, as life scientists would come to call it, gradually abandoned anthropometry as a core methodology while biomedicine, as a 'new way of knowing' (Pickstone 2000), brought dramatic changes, for example in the scale of investment in research, with regard to the state as a scientific entrepreneur, and through a closer relationship between the laboratory and the clinic (Quirke and Gaudillière 2008). Although the new molecular approach in the life sciences was already underway, none of the chapters in this volume deals with a colonial context in which molecular biology would already have been applied by that time. This volume, then, takes the commonly accepted chronology of the emergence of biomedicine as a temporal frame and examines the wide range of issues that were analysed and administered under the purview of health and difference. The chapters connect and highlight population level themes related to health that colonial administrators and life scientists concerned themselves with: a population's vitality, with special regard to population 'mixtures', labour, infectious and chronic diseases, nutrition, and reproduction.

What does the concert of the chapters in this volume make audible? Or, what is colonial about medicine and Western thinking on human difference everywhere? Across the chapters, one can perceive a common thread in the activities of administrators and researchers. They were engaged in risky and gigantic projects that entailed – or hoped to entail – the management of entire populations.[2] Often enough, the main driving force in these management activities was pecunial exploitation or the maximization of territorial gains and military power. As well, and often relatedly, managing populations meant contending with medical and hygienic concerns and the logistics of food supplies. What made this management so perfect a context for collaboration between administrations and scientists were the parameters that needed to be managed. Most of these parameters could be expressed in quantitative terms. They concerned the management of proximity and distance, both physical and social. Their mandates included handling increases or scarcities, of food supplies but also of population numbers, or of available workers. Furthermore, it involved managing the exposure and protection of certain populations, sometimes with the perception that populations had specific abilities to withstand their circumstances, like, for example hard labour or lower calorie diets. If managed optimally, these relations promised to guarantee a smooth process of adding value. Health, food, feelings, animals, plants, landscape, working places, settlements etc. were all meant to be aligned to this process that required so much effort to manage. In addition to the material dimensions, there was also symbolic value to be considered: managing increases in value was also about demonstrating to the world that a colonial power was able to bring life to a population and to turn it into a thriving productive one – a 'civilized' one. Managing specificity was about efficiency, but it was also about proudly presenting healthy, strong, disciplined, willing individuals.

This project of making differences a viable tool for management points to a tension that has been aptly described by Bowker and Star: 'Classification and standards are two sides of the same coin'. And yet, they state, 'classifications may or may not become standardized [...] while every successful standard imposes a classification system, at the very least between good and bad ways of organizing actions or things' (2000: 15). Bowker and Star also emphasize the role of classifications and standards as 'objects for cooperation across social worlds' (ibid.: 283), or boundary objects. As, for example, Letícia Galluzzi Bizzo's chapter shows, establishing an international standard for nutritional requirements became complicated when new categories needed to be considered in the context of global cooperation for nutritional standards.

Writing accounts of colonial medicine and studies of human difference entails challenges in how to use historical actors' terms, like 'race', 'blacks', 'civilized', 'natives'. These terms have despicable histories and prejudicial meanings that we clearly do not want to replicate in the present. And yet, citing the terms cannot be avoided because they were also common descriptions of categories of people at the time. Furthermore, the terms are also used today with different sensibilities depending on the context. To deal with this situation, however imperfectly, wherever possible the contributors have indicated that such terms are historically accurate actors' terms by using quotation marks. The contributors work to contextualize them in the conventions of the day.

The chapters deal with the management, administration and study of the following health concerns: a population's overall health and vitality, labour, infectious and chronic diseases, nutrition and reproduction. These are issues that today are increasingly considered to be biosocial or biocultural. By drawing on the chapters of this volume, we aim to demonstrate how each of those concerns led to both differentiation and standardization, as two sides of the same coin, both of populations and of health care solutions (Bowker and Star 2000: 15).

Health And Vitality

The very definition of health was researched with respect to a population's biological and social characteristics. Such lines of inquiry were connected to colonial concerns such as indigenous populations' suitability for a labour force or Europeans' ability to live outside Europe. Cândido da Silva's chapter presents a German tropical medicine expedition to Espírito Santo, Brazil in 1936, when Germany had aspirations of regaining colonies. The team leaders, Giemsa and Nauck, studied three generations of German settlers by taking anthropometric measurements and genealogies. They also compared the heights and weights of school children in Germany with third-generation German children in Espírito Santo. Differences in growth rates were explained not by genetic inheritance, but by phenotypical aspects, and exposure to tropical conditions in particular. From this, Cândido da Silva argues that although the researchers, Giemsa and Nauck, did engage with the racial thinking popular at that time in National Socialist Germany, they ultimately put more emphasis on social and environmental dimensions. In their studies, the Germans in Espírito Santo, Brazil were not so much 'racially degenerate' as unhealthy and isolated.

'Race' was an explanation for both the presence and absence of the vigour of the indigenous population on Nauru, a Pacific island administered by the German Pacific Phosphate Company from 1906 to 1919. German colonial health reports and the Pacific Phosphate Company's correspondence reveal that prior to the company's need for labour, researchers perceived Nauruans as a happy, mentally and physically healthy and vigorous 'race'. With the introduction of waged labour in the phosphate mines, however, when the Nauruans showed no interest in mining and contracted introduced diseases, they became a weak 'race', ill-suited for strenuous work.

A population's vitality was often threatened by infectious or endemic diseases, but as Jean-Paul Bado's chapter shows, this emphasis led early twentieth-century colonial physicians to overlook the presence of chronic or non-infectious diseases like cancer. In his consideration of primary liver cancer (PLC) in Africa, Bado shows how this oversight was connected to perceptions of difference: cancer was a disease of European modernity, and since African lifestyles were perceived as 'natural' or 'uncivilized', cancer was underdiagnosed in the first half of the twentieth century. As diagnostic technologies became available, the evidence no longer supported the earlier colonial ideology and the methodologies of comparing 'civilized' with 'bon sauvage' could be abandoned, but a significant number of doctors and researchers were slow to change. By the 1950s, researchers were assessing ecological factors present in many French regions, including the role of food or rations, endemic parasites, and bacterial and viral infections. Eventually they were able to rule out ecological factors, and settled on aflatoxin, a toxin produced by the storage methods of peanuts, maize and many other grains, as the cause.

Labour

Having a healthy labour force was a common demand from colonial settlers and a running administrative concern. Studying the health and suitability of potential workers, as Sarah Ehlers, Jean Mitchell and Antje Kühnast's chapters show, resulted in situations where implicitly or explicitly, researchers, medical personnel or administrators dealt with or explained the health issue by comparing population categories, such as 'race', other biological factors, and culture. Studies of (potential) labourers entailed the study of infectious diseases, like malaria (Mitchell) or sleeping sickness (Ehlers). Ehlers, using internal correspondence between doctors, researchers and bureaucrats, discusses how in British Uganda, research practices and interventions on sleeping sickness varied according to the patient's 'race'. She writes that for Africans, doctors would travel to the infected areas to assess the epidemic. The doctors would then detain the sick in sleeping sickness camps, where they could be studied and offered treatments. The colonial doctors were not only interested in the infected individuals but they also attempted to assess the patient inmates' social structure, such as the names and size of villages and their subsistence practices. For Europeans, sleeping sickness was presented not in broad reports, but as individual case studies, indicating that the disease was seen as a threat to individual settlers. In terms of treatment, given sufficient resources, the same doctors would treat Europeans at home or in a European hospital and treat Africans in a sleeping sickness camp.

In the German phosphate company's administration of Nauru in the south Pacific, Kühnast demonstrates, the inadequate handling of epidemics among the Nauruans and the Chinese was not due to capitalist extraction of labour alone. The company's failure to curb epidemics was connected to perceptions of cultural, mental and physical differences – buttressed by researchers – that explained the incompatibility of certain human groups for particular kinds of work and lifestyles.

Nutrition

Nutritional sciences were part of the biopolitical strategies of nation states, but were also central concerns for international organizations, like the Health Organisation of the League of Nations (HOLN), and later the Food and Agriculture Organisation of the United Nations (FAO). Writing of the first half of the twentieth century, Letícia Galluzzi Bizzo demonstrates how the scientific conventions aimed to measure universal human nutrient requirements in the form of an index of daily caloric values. These were used in international organizations' administrative aspirations to monitor world food circumstances and reduce the social problems of hunger around the world. Already then, life scientists recognized the role of living conditions for human body variation, which included labour, diet and social inequalities. During the global depression, the health of a labour force became a pressing global problem. Two renowned experts with colonial experience, Wallace Aykroyd and Gerard van Veen André, were recruited whose publications would have enduring influence at the UN. Their methodologies generated population categories, like 'rice eaters', from central foods in a population's diet and the knowledge that rice based diets appeared to provide sustenance with fewer calories. A complex negotiation between UN subcommittees, advisory groups and inter-governmental conferences resulted in differential calorie allowances for non-Western people, extrapolated from the so-called 'Rice eaters' in the publications.

Barbara Cooper focuses on a nutritional science research endeavour, the 'Mission Anthropologique' in French West Africa's colonies, which aimed at a general understanding of malnutrition, diets, food requirements, labour capacities and population specificities. The researchers took a particular interest in the situation in Niger in the late colonial period, where the problem of malnutrition had historically been acute. Cooper highlights the enduring significance of the military and highly masculine character of French colonial health services, and how this led to an imbalanced consideration of women and men in the survey. Her chapter reminds us that the reports and documentations of the world on paper portrayed a situation dramatically different from the realities on the ground, at least in those areas that were far removed from cities. Curiously, the Mission Anthropologique relied heavily on anthropometric surveys, in addition to a stunning variety of disciplines, and in practice it took the robust, well-fed soldier as the ideal norm for the whole population. The dramatic malnutrition that the researchers nevertheless observed was blamed on the indigenous populations, on their poor agricultural knowledge and, paradoxically, on their ignorance of traditional food practices that were rich and healthy even in the eyes of those Western scientists. Hence, they frequently resorted to deterministic explanations based upon 'race' and later 'culture' to account for debilitating malnutrition in the absence of sufficient resources to adequately address the problem.

Reproduction

Attempting to understand how populations, or portions of populations, grew or declined was also a central colonial anxiety and scientific preoccupation. In addition to labour migration, particularly in the first decades of the twentieth century, researchers' explanations for unusual reproductive patterns were often framed in terms of the biological and cultural differences of particular populations. The scientific questions of population growth or decline were frequently connected to colonial endeavours to regulate aspects of what they deemed part of reproduction such as marriage practices (e.g. Widmer 2014), birth attendants (e.g. Thomas 2003; Boddy 2007) or sexuality (e.g. Stoler 2002). Connected to these themes, especially in populations that were in dramatic decline, explanations for infant mortality figured prominently. The poor nutritional status of infants, as Cooper's chapter demonstrates, was understood through research on cultural differences in mothering and mothers' feeding practices (this was also the case in other regions of the world, e.g. Jolly 1998; McElhinny 2005). In the AOF, Cooper argues, the emphasis on cultural differences meant that the political economic reasons that caused hunger and malnutrition could be overlooked.

Trying to understand differences in infant mortality was dependent on population-wide information. However, in many colonial situations this information was in short supply. Only in the case of Nauruans, examined by Kühnast, did research conditions seem to be perfect: researchers believed they could easily track the thousand Nauruans living in the territory. But as Samuel Coghe describes in his chapter, because of the lack of census and vital statistics, colonial physicians in Portuguese Angola had to count people themselves as an important aspect of their jobs as physicians. It was the *indigenas* whose populations were of concern due to 'depopulation' and whose difference was believed to be based on their biology and relationship to 'civilization'. The physician's presumptions about fertility and mortality patterns mirrored differences between Africans and Europeans as their healthcare interventions, such as they were, were exclusively directed at the *indígenas* ('natives'). Indeed, the settler populations were not studied in the same way. Coghe examines how these physicians, who became colonial demographers by necessity, viewed, interpreted and used their data. By addressing the particularly close connection between medicine and demography in interwar Angola, his chapter demonstrates the links between governance, health and the categorization of difference within populations as part and parcel of colonial practices.

Intra-Population Differences

Beyond the obvious hierarchical relationships between colonizer and subject population forged in social, political and material inequalities, certain chapters also ask what other kinds of categories and levels of differentiation were highlighted as more or less relevant in the administration and research on health in colonial contexts. Researchers relied on differences within populations in the same territory as variables in their explanations for colonial problems. Importing indentured labour could, for example, bring different colonial subjects into contact with one another. This would have administrative implications for officials in managing different population segments, as well as research opportunities for experts to weigh in on how to manage population differences between, for example, 'traditional' people and imported labourers. In the colonial southwestern Pacific, for example, the depopulation of indigenous Islanders was a scientific and administrative problem. Traditional culture was operationalized as a variable in research methods and explanations for poor health. In Fiji, British administrators applied different logics of state control over Indian indentured labour on plantations, and indigenous Fijians living in traditional villages (Thomas 1990). In the New Hebrides, as Widmer (2012) writes and Mitchell reiterates here, health researchers in the 1920s cautioned the British against 'hybridization' between indentured labourers and New Hebrideans if Indian labourers were imported. Indeed, the Tonkinese (from the gulf of Tonkin in French Indochina, now Vietnam) plantation labourers, indentured by the French, lived in labour camps often quite separate from indigenous villages in the New Hebrides. Cooper writes that colonial nutrition sciences in the AOF classified bodies according to their relation with traditional culture, leading to gendered kinds of interventions. For example, nutritional deficiencies, in particular regarding children, were blamed on women's traditional feeding styles and weaning practices, both part of their 'culture', rather than economic inequality. In Portuguese Angola, as Coghe writes, medical demographers blamed cultural factors for some aspects of population decline, while also using some racial categorizations to delineate populations of Bantu and 'Bushmen'.

Animals, Insects and Environments

Colonial administrators and life scientists also had some shared concerns in understanding the relationships that particular 'races' or segments of populations had with environments, animals, climates and 'nature' (e.g. Haraway 1989; Macleod 2000; Arnold 2006; Deb Roy 2013). In this respect, additional sets of categorizations beyond colonizer/colonized were highlighted. Recognizing that the disease etiologies often connected humans to broader ecologies, the researchers applied categories of difference to the human populations to understand differential rates of illnesses (Lipphardt 2016). Mitchell documents how entomologists, concerned with eliminating malaria in the southwestern Pacific, were intensely interested in human relationships with mosquitos. During the Second World War, researchers measured flights of mosquitos in order to know how far apart soldiers' camps would have to be from 'natives', who were considered to have different vulnerabilities to malaria than the American troops. In trying to solve the malaria problem, made acute by the Pacific theatre of the war, Mitchell argues, human-insect relations were racialized. There were similar strategies in changing relationships with the landscape and the tsetse fly, as Ehlers explores, in the prevention and containment measures enacted during the sleeping sickness campaigns in the British Protectorate of Uganda. With this, the volume is less concerned with how ecology was part of racial thought in the first part of the twentieth century (which has been well documented by, for example, Anderson 2003), but rather with how colonial interventions on landscapes were connected to the administration of diverse populations and their health.

Colonial and Scientific Critiques

Scientific knowledge about human variation was not limited to circuits between Europe and its locations of colonial control. Scientific methods could also be turned around and used to critique the colonial population categories themselves.[4] The Indo-European social scientist at the centre of Hans Pols' chapter, Joseph Koks, provides a wonderfully satirical account of the absurdities of colonial and scientific classifications. Pols shows how administrative and daily-life classifications differed in the Dutch East Indies, and reminds us that what became documented in colonial administrative practises was by no means congruent with common sense. Koks complained that anthropological conventions of defining populations by 'race' distinguished between 'Europeans', 'Indo-Europeans' and 'natives', while administrative practises only distinguished between 'Europeans' and 'natives'. In effect, as he argued in a painstaking account of his complex social reality, these classifications denied that 'Indo-Europeans' had crucially contributed to the accomplishments that were all ascribed to the Europeans. To be 'Indo-European' was to be poor, in the social reality of Batavia. But, if counted according to the (physical) anthropological definition, and if the role of social class was ignored in the categorization, he claimed that ninety per cent of all Europeans in the colony were to be ascribed to the Indo-European category. While Koks employed race thinking to expose the social prejudices that prevented the recognition of the achievements of 'Indo-Europeans', he also, rather presciently, critiqued physical anthropologists' overdetermined use of race concepts to explain broad social issues: 'Imagine a Chinese or African anthropologist, who explains the crisis [the Great Depression] by referring to the nature of the white race!'

This collection of chapters covers a wide range of administrative and scientific concerns in differing human populations and imperial situations. It offers, on the one hand, a dynamic context to think about relationships between scientific categories used to identify and classify human populations. These chapters also show how differences within populations and between populations have to be understood as historical relationships in concert with their scientific treatment as disciplinary objects or tools for understanding health or making scientific claims about human variation.

Conclusion

What is the postcolonial present that we are writing the history of, with respect to the life sciences, administration of health and notions of human difference? The social forms of colonial routines and the historically shaped relationships between populations forged in colonial times have remarkably long half-lives. How different are the contemporary short tours of expert consultants to gather information from the colonial patrols of decades ago? The avalanche of reports, commissions of inquiry, memos aimed at developing (or improving) the health of people in the global south did not stop with the formal end of colonialism but continue to be instruments of governmentality, if not improved health conditions. The colonial connections between categorization, classification and governance now find their way to being relevant 'indicators' (Merry 2011) in 'global health'. This is but one aspect of the continuing postcolonial need to keep the colonial legacies in central analytical focus, Anderson (2014) reminds us, when analysing 'global' aspects of the circulations of biomedical knowledge and practice.

In recent years there has been a marked increase in the speed and scope of technologies to analyse the genetic material of individuals and populations. This has purportedly had far reaching implications for individual and population health, not least of which is how global health problems and solutions are framed. Accordingly, social scientists have been increasingly interested in engaging with these developments and what they mean for health and what it means to be human. Social scientists' concern for the social implications of genetic technologies interrogate the deterministic dimensions implied in divisions between nature and culture, or nature and nurture, to think rather in terms like 'biosocial', 'local biologies' or 'naturecultures' (e.g. Lock and Kaufert 2001; Haraway 2003; Ingold and Palsson 2013). This volume can be seen as providing a historical context for these scholarly engagements with scientists' work on categorizing human populations. What these chapters contribute is a heightened sensitivity towards the very local and contingent nature of scientific *and* administrative categories, of their emergence, usage and circulation and colonial context (cf. Reardon 2005; Kowal, Radin and Reardon 2013; Wade et al. 2014). These insights are relevant, for example, for how we might view the categories of difference applied in the contemporary construction and usage of any genomic database used in the service of genetic explanations for global health issues. Furthermore, the same sensitivity should be in place when it comes to considering deterministic claims and outcomes of any classification, be they biological, genetic, social or cultural. The chapters in this volume demonstrate that a wide range of perceptions of populations' characteristics as 'innate' or malleable was possible in the minds and practices of those who were to decide upon the lives of many. It is important to consider these differentiations – fixed or changeable – in order to understand why certain actions were taken with sometimes seemingly contradictory justifications. Would actors have agreed on the malleability of the characteristics they ascribed to a certain group? On what level, if at all, would they have claimed that such characteristics were (not) changeable? Or would they have assumed malleability only within a certain time frame? If not in an individual's life span, then maybe for their children? Within a few generations? Or only after dozens of generations, with evolutionary change? What would the assumed mechanism of change be? Self-discipline? Education? Exposure? Or by natural selection? These questions are still potent when it comes to the question of how and why populations and individuals differ, and how distinctive they will supposedly remain in the future. The stakes were (and are) high: this knowledge pertained, after all, to lofty questions of what it meant to be human and what our relationship with other organisms on the planet should and can be.

Finally, we would like to take up Anderson's challenge once again. Was there something colonial about Western medicine and about categorizing human variation everywhere? The chapters in this volume suggest that attempts to research human variation represented efforts to make human difference productive for improving and governing human health. Differences were put to work to understand and administer populations' vitality, food requirements, reproduction and labour. A population's chronic or infectious diseases, seen as connected to agricultural and environmental specificities, were impossible to understand and administer without recourse to categories of human difference.

Cutting through many of the chapters is the colonial recognition that poverty or social inequality was a cause of poor health: an acknowledgement that continues to be heard in social policy and research in the present day. Yet, as Galluzzi Bizzo puts it, researchers and administrators found ways of 'postponing equality'. This, then, points to the fact that even if poor health was understood to be a consequence of inequality, in the end those involved shied away from finding adequate answers to the most pressing problems of humanity. This volume combines the history of medicine and the history of the life sciences, by focusing on the colonial aspects of both, and particularly on their shared use of the biological and social categories of human variation, during the first half of the twentieth century. The chapters clearly demonstrate how the historical emergence of biomedicine as a global social practice, together with scientific accounts of human variation, went hand in hand with the social and material life of accountancy and governance in colonial rule.

Alexandra Widmer is an anthropologist with expertise in colonial and postcolonial population politics and health in the Pacific Islands. Her publications include *The Imbalanced Sex-Ratio and the High Bride Price: Watermarks of Race in Demography and the Colonial Regulation of Reproduction* (2014) and *Diversity as Valued and Troubled: Social Identities and Demographic Categories in Understandings of Rapid Urban Growth in Vanuatu* (2013). She held a research scholar position at the Max Planck Institute for the History of Science and now teaches in the Anthropology department at York University in Toronto.

Veronika Lipphardt works on the history of the life sciences, particularly physical anthropology and human population genetics in their political, social and cultural contexts. Since 2015, she has been professor for STS at the University College

Freiburg. She was the director of an Independent Research Group at the Max Planck Institute for the History of Science. In 2008 she published a book about German-Jewish scientists and their contribution to the scientific debate about the so-called 'Jewish race' between 1900 and 1935. Her forthcoming book will deal with the history of knowledge about human variation in the twentieth century.

Notes

Authors are listed in alphabetical order, reflecting equal contributions.

1. See also Solomos and Back (1994) and Back and Solomos (2009).
2. For population research and management in colonial contexts, see for example Hartmann and Unger (2014), Bashford (2014) and Deb Roy (2015).
3. But also apart from explicit critiques, what Tilley calls 'endogenous idioms of group difference and rank' deserves attention: the categories the colonized subjects would call upon, most notably in their efforts to resist colonial rule (see Tilley 2014: 779).

Bibliography

Anderson, W. 1998. 'Where is the Post-Colonial History of Medicine?', *Bulletin of the History of Medicine* 72(3): 522–530.

—— 2003 'The Natures of Culture: Environment and Race in the Colonial Tropics', in P. Greenough and A. Tsing (eds), *Nature in the Global South: Environmental Projects in South and Southeast Asia*. Durham: Duke University Press, pp. 29–46.

—— 2006. *Colonial Pathologies: American Tropical Medicine, Race and Hygiene in the Philippines*. Durham: Duke University Press.

—— 2009. 'Ambiguities of Race: Science on the Reproductive Frontiers of Australia and the Pacific between Wars', *Australian Historical Studies* 40(2): 143–160.

—— 2014. 'Making Global Health History: The Postcolonial Worldliness of Biomedicine', *Social History of Medicine* 27(2): 372–384.

Appadurai, A. 1993. 'Number in the Colonial Imagination', in C.A. Breckenridge and P. van der Veer (eds), *Orientalism and the Post-Colonial Predicament*. Philadelphia: University of Pennsylvania Press, pp. 314–339.

Arnold, D. 1993. *Colonizing the Body: State Medicine and Epidemic Disease in Nineteenth Century India*. London: University of California.

—— 2006. *The Tropics and the Travelling Gaze: India, Landscape and Science, 1800-1856*. Seattle: University of Washington Press.

Back, L. and J. Solomos (eds). 2009. *Theories of Race and Racism*. London: Routledge.

Barkan, E. 1992. *The Retreat of Scientific Racism: Changing Concepts of Race in Britain and the United States between the Two World Wars*. Cambridge: Cambridge University Press.

Bashford, A. 2014. *Global Population: History, Geopolitics, and Life on Earth*. New York: Columbian University Press.

Bayly, C. 1996. *Empire and Information: Intelligence Gathering and Social Communication in India, 1780–1870*. Cambridge: Cambridge University Press.

Boddy, J. 2007. *Civilizing Women: British Crusades in Colonial Sudan*. Princeton/Woodstock: Princeton University Press.

Boris, E. and A. Janssens. 1999. 'Complicating Categories: an Introduction', *International Review of Social History* 44(S7): 1–13.

Bowker, G.C. and S.L. Star. 2000. *Sorting Things Out. Classifications and Consequences.* Cambridge: MIT Press.

Cohn, B.S. 1996. *Colonialism and its Forms of Knowledge. The British in India.* Princeton: Princeton University Press.

Deb Roy, R. 2013. 'Quinine, Mosquitoes and Empire: Reassembling Malaria in British India, 1890-1910', *South Asian History and Culture* 4(1): 65–86.

——— 2015. 'Nonhuman Empires', *Comparative Studies of South Asia, Africa and the Middle East* 35(1): 66–75.

Dirks, N. 2004. 'The Ethnographic State', in S. Dube and S. Amin (eds), *Postcolonial Passages: Contemporary History-Writing on India.* New Delhi: Oxford University Press, pp. 70–88.

Douglas, B. and C. Ballard (eds). 2008. *Foreign Bodies: Oceania and the Science of Race 1750-1940.* Canberra: Australian National University Press.

Douglas, M. and D. Hull. 1992. *How Classification Works: Nelson Goodman among the Social Sciences.* Edinburgh: Edinburgh University Press.

Dubow, A. 2006. *A Commonwealth of Knowledge: Science, Sensibility, and White South Africa, 1820-2000.* Oxford: Oxford University Press.

Ernst, W. 1999. 'Historical and Contemporary Perspectives on Race, Science and Medicine', in W. Ernst and B. Harris (eds), *Race, Science and Medicine, 1700-1960.* London: Routledge, pp. 1–29.

Ernst, W. and B. Harris (eds). 1999. *Race, Science and Medicine, 1700-1960.* London: Routledge.

Fleck, L., T.J. Trenn and R.K. Merton (eds). 1979. *Genesis and Development of a Scientific Fact.* Chicago: University of Chicago Press.

Haraway, D. 1989. *Primate Visions: Gender, Race and Nature in the World of Modern Science.* London: Verso.

——— 2003. *The Companion Species Manifesto: Dogs, People, and Significant Otherness.* Chicago: Prickly Paradigm Press.

Harrison, M. 1999. *Climates and Constitutions: Health, Race, Environment and British Imperialism in India, 1600-1850.* New Delhi: Oxford University Press.

Hartmann, H. and C.R. Unger (eds). 2014. *A World of Populations. Transnational Perspectives on Demography in the Twentieth Century.* Oxford: Berghahn Books.

Hawkins, S. 2002. *Writing and Colonialism in Northern Ghana: The Encounter between the LoDagaa and the 'World on Paper'.* Toronto: University of Toronto Press.

Ingold, T. and G. Palsson (eds). 2013. *Biosocial Becomings: Integrating Social and Biological Anthropology.* Cambridge: Cambridge University Press.

Jolly, M. 1998. 'Other Mothers: Maternal "Insouciance" and the Depopulation Debate in Fiji and Vanuatu, 1890-1930', in K. Ram and M. Jolly (eds), *Maternities and Modernities: Colonial and Postcolonial Experiences in Asia and the Pacific.* Cambridge: Cambridge University Press, pp. 177–212.

Kowal, E., J. Radin and J. Reardon. 2013. 'Indigenous Body Parts, Mutating Temporalities, and the Half-lives of Postcolonial Technoscience', *Social Studies of Science* 43(4): 465–483.

Lipphardt, V. 2012. 'Isolates and Crosses in Human Population Genetics; or, a Contextualization of German Race Science', *Current Anthropology* 53(S5): 69–82.

——— 2014. '"Geographical Distribution Patterns of Various Genes": Genetic Studies of Human Variation after 1945', *Studies in History and Philosophy of Biological and Biomedical Sciences* 47: 50–61.

——— 2016. '"Europeans" and "Whites": Biomedical Knowledge about the "European Race" in Early Twentieth Century Colonial Contexts', *Comparativ* 25 (5-6): 137–146.

Lock, M. and P. Kaufert. 2001. 'Menopause, Local Biologies, and Cultures of Aging', *American Journal of Human Biology* 13(4): 429–567.

McElhinny, B. 2005. '"Kissing a Baby Is Not at All Good for Him": Infant Mortality, Medicine, and Colonial Modernity in the US-Occupied Philippines', *American Anthropologist* 107(2): 183–194.

Macleod, R. 2000. 'Nature and Empire: Science and the Colonial Enterprise', *Osiris* 15: 1–13.

Merry, S.E. 2011. 'Measuring the World', *Current Anthropology* 52(S3): 83–95.

Osborne, M.A. 2005. 'Science and the French Empire', *Isis* 96(1): 80–87.

Pande, I. 2010. *Medicine, Race and Liberalism: Symptoms of Empire*. New York: Routledge.

Pels, P. 2008. 'What has Anthropology Learned from the Anthropology of Colonialism?', *Social Anthropology* 16(3): 280.

Perez Hattori, A. 2004. *Colonial Dis-Ease: US Navy Policies and the Chamorros of Guam, 1898-1941*. Honolulu: University of Hawaii Press.

Pickstone, J.V. 2000. *Ways of Knowing: a New History of Science, Technology and Medicine*. Manchester: Manchester University Press.

Prakash, G. 1999. *Another Reason: Science and the Imagination of Modern India*. New Jersey: Princeton University Press.

Quirke, V. and J.-P. Gaudillière. 2008. 'The Era of Biomedicine: Science, Medicine, and Public Health in Britain and France after the Second World War', *Medical History* 52(4): 441–452.

Reardon, J. 2005. *Race to the Finish: Identity and Governance in an Age of Genomics*. Princeton: Princeton University Press.

Rich, P. 1990. 'Race, Science, and the Legitimization of White Supremacy in Modern South Africa, 1902-1940', *International Journal of African Historical Studies* 23: 665–686.

Roque, R. 2010. *Headhunting and Colonialism. Anthropology and the Circulation of Human Skulls in the Portuguese Empire, 1870-1930*. New York: Palgrave Macmillan.

Roque, R. and K. Wagner (eds). 2012. *Engaging Colonial Knowledge: Reading European Archives in World History*. London: Palgrave MacMillan.

Scott, D. 1995. 'Colonial Governmentality', *Social Text* 43(3): 191–220.

Solomos, J. and L. Back. 1994. 'Conceptualising Racisms: Social Theory, Politics and Research', *Sociology* 28(1): 143–161.

Stepan, N. 1982. *The Idea of Race in Science: Great Britain 1800-1960*. Hamden: Archon.

Stoler, A.L. 1995. *Race and the Education of Desire: Foucault's History of Sexuality and the Colonial Order of Things*. Durham/London: Duke University Press.

——— 2002. *Carnal Knowledge and Imperial Power: Race and the Intimate in Colonial Rule*. Los Angeles: University of California Press.

——— 2009. *Along the Archival Grain Epistemic Anxieties and Colonial Common Sense*. Princeton: Princeton University Press.

Thomas, N. 1990. 'Sanitation and Seeing: The Creation of State Power in Early Colonial Fiji', *Comparative Studies in Society and History* 32(1): 149–170.

——— 2003. *Politics of the Womb: Women, Reproduction, and the State in Kenya*. Berkeley: University of California Press.

Tilley, H. 2011. *Africa as a Living Laboratory: Empire, Development, and the Problem of Scientific Knowledge, 1870-1950*. Chicago: Chicago University Press.

——— 2014. 'Racial Science, Geopolitics, and Empires: Paradoxes of Power', *Isis* 105(4): 773–781.

Vaughan, M. 1991. *Curing their Ills: Colonial Power and African Illness*. Stanford: Stanford University Press.

Wade, P., C. López Beltrán, E. Restrepo and R. Ventura Santos. 2014. *Mestizo Genomics: Race Mixture, Nation, and Science in Latin America.* Durham: Duke University Press.

Weindling, P. 1993. *Health, Race and German Politics between National Unification and Nazism, 1870-1945.* Cambridge: Cambridge University Press.

Widmer, A. 2008. 'The Effects of Elusive Knowledge: Census, Health Laws and Inconsistently Modern Subjects in Early Colonial Vanuatu', *Journal of Legal Anthropology* 1(1): 92–116.

——— 2012. 'From Research Encounters to Metropolitan Debates: The Making and Meaning of the Melanesian "Race" during Demographic Decline', *Paideuma* 58: 69–93.

——— 2014. 'The Imbalanced Sex Ratio and the High Bride Price Watermarks of Race in Demography, Census, and the Colonial Regulation of Reproduction', *Science, Technology & Human Values* 39(4): 538–560.

Zimmerman, A. 2001. *Anthropology and Anti-humanism in Imperial Germany.* Chicago/London: University of Chicago Press.

1

Race, Health and Colonial Politics in the Third Reich

Nauck and Giemsa's Expedition to Espírito Santo, Brazil in 1936

André Felipe Cândido da Silva

From March to May 1936, Ernst Nauck and Gustav Giemsa, researchers at the Hamburg Institute for Maritime and Tropical Diseases, visited a region of Espirito Santo – a southeastern state of Brazil – that had been settled by German immigrants. The aim of the expedition was to investigate the health and living conditions of this population group in order to determine whether – and to what extent – they had 'acclimatized', i.e. whether they had adapted to the area while maintaining their typically German 'racial' features, or whether they had 'degenerated' in the tropics, having suffered negative changes due to the tropical environment.

The objective of this chapter is to analyse Nauck and Giemsa's trip to the state of Espírito Santo in the context of tropical medicine in the Third Reich and its close association with the colonialist movement of the period. I suggest that their arguments are consistent with the prevailing intellectual environment in tropical medicine during the Third Reich, which was marked by the ideas of racial hygiene, the ambitions of the colonial movement, institutional disputes and theoretical controversies.

Although the region visited by Nauck and Giemsa was not a target of the colonialist ambitions of National Socialism, the tropical medicine specialists from the Hamburg Institute gave meaning to the trip by their intention to participate in colonial policy and to take the lead in the public health management of the colonial territories of Africa as soon as they were reconquered. In this sense, the trip was linked to the strong colonial revisionism within the Hamburg Institute, which began with the loss of the colonies after the Versailles Treaty, as shown by Stefan Wulf (1994), but which assumed new meaning during the National Socialist regime, making it an object of controversy that divided tropical medicine at the time.[1] The visit to Espírito Santo was connected to the interests of tropical physicians and other sectors involved in politics and the National Socialist colonial movement, that is, how their research 'was connected to colonial administrative requirements', the focus – as Widmer and Lipphardt assert – of this book, even though we are dealing with territories and populations not subject to formal colonialism and with a colonial project that failed to materialize. To what extent does a voyage of this type and the conclusions drawn from it contribute to the understanding of the colonial administration projects then planned by National Socialist Germany? How did the administrative practices proposed – but never put into effect due to the fate of the German armies in the Second World War – intend to deal with human diversity in African territories?

In this sense, we must analyse which tools were employed by Nauck and Giemsa to study the populations of German origin in Brazil. From what records, practices and approaches did they build their discourse on the colonial groups they visited? How did their scientific practices put them 'alongside ideologies, discourses and knowledge claims', as the chapters in this book intend to demonstrate?

On the one hand, Nauck and Giemsa began with the fixed, unchangeable idea of a 'German race', but at the same time they and other tropical physicians accepted the possibility of transformation due to climatic factors. In a sense, this brings to mind what Alexandra Widmer claimed to be 'the malleability of racial thinking and the intransigence of racial categorization' (2012: 69). In the end, the goal of the studies was to determine which characteristics of the 'German race' were immutable and hereditary and which were subject to variations due to environmental factors of natural and socio-economic types. Both researchers dealt with diversity that they recognized as arising less from physical and biological factors than as a result of socio-cultural factors, such as geographical and cultural isolation, economic, health and lifestyle habits. Reflecting the medical and anthropological thinking of the period, they condemned racial miscegenation. They identified this as the greatest threat to their 'Germanness' (*Deutschtum*) both in terms of 'racial' constitution and maintenance of the habits that identified them as authentically German. Thus, the findings of Nauck and Giemsa were in line with the strict racial segregation laws that Germany intended to impose on African territories after occupation by the Germans (Eckart 1997).

In addition to explaining the connection between the trip to Espírito Santo and National Socialist colonial policy and how the tropical physicians allied themselves with it, I intend to explore how Nauck and Giemsa's observations dialogued with the concepts of human diversity and heredity then prevailing in the Third Reich, when the idea of race became central in addressing the difference between social formations, cultural expression, behaviour, populations and perceived health and disease.[2] I argue that the findings of the two investigators on racial acclimatization are marked by strong ambivalence, a result of the way in which they appropriated ideas on race while remaining faithful to a 'thinking style' (Fleck 1979) focused on public health practices and the possibility of settling Europeans in the tropics, once the issue of 'tropical diseases' was addressed. This ambivalence is less a weakness or inconsistency than a feature of racial discourse that, according to Ernst, 'works well not despite its logical inconsistencies, ambiguities and mixing up of premises, but because of them' (1999: 1–28).

This analysis of Nauck and Giemsa's trip to Espírito Santo was based on the dense report published in 1939 in the Annals of the Hanseatic University (the name given to the University of Hamburg by the National Socialists) and the researchers' other publications. Another important reference is the report by Ernst Wagemann, who visited the same region in Espírito Santo in 1912. His findings, published in 1915, formed the backbone of Giemsa and Nauck's work, and they, comparing their findings to his, then expanded on and updated them (Giemsa and Nauck 1939).[3]

The Historical Archives of the Bernhard Nocht Institute for Tropical Medicine (*Bernhard Nocht Institut für Tropenmedizin*), as the Hamburg Institute for Maritime and Tropical Diseases is now called, hold the only vestige of Nauck and Giemsa's trip to Espírito Santo in the form of a typed version of the text that was published in 1939, in addition to other publications located in the reprint collection. Ernst Nauck's papers in the Hamburg City Archive show no sign of sources relating to his time in Brazil in 1936. Therefore, I did not have access to the records of observations, protocols and notes taken during the visit to the Espírito Santo colonies. The report is terse with respect to the specific methods they adopted in their studies. It mentions only that they took anthropometric measurements, but does not describe which protocols and theoretical frameworks they were based on.

The population data upon which Nauck and Giemsa based their claim of a satisfactory population growth rate and low mortality index is based on information from Wagemann and updates to it and on records kept by the Catholic and Lutheran churches. The extensive nature of the religious organizations was fundamental not only for providing the researchers with access to the settlers, but also for supplying the data that supported the studies. They had also assisted Ernst Wagemann in 1912, and he recognized the value of these sources: 'The ecclesiastical books are an invaluable source. They kept records almost from the beginning of colonization, in all communities, with care worthy of admiration' (1915).[4] Interestingly, the ecclesiastical records were also sources for studies that entomologist Patrick A. Buxton carried out in the New Hebrides islands to assess the incidence of malaria among children, as Jean Mitchell mentions in Chapter 3 of this book.

Although Giemsa and Nauck did address the ideas of racial hygiene touted in National Socialist Germany, they ultimately put more emphasis on factors of a social and cultural order in their portrayals of the German communities in Espírito Santo. In their eyes, they were not so much racially degenerate as simply unhealthy, and culturally and geographically isolated.

The German settlements in Espírito Santo

Why did Nauck and Giemsa choose Espírito Santo, in southeastern Brazil, for their study? They justified their interest in Espírito Santo because, in their view, those settlements were amongst the few population groups that had remained 'racially pure' for around three generations, and had also maintained their cultural characteristics: 'The regions in which a population group from Europe has settled for many generations without mixing with the natives are of the greatest interest … In the Brazilian state of Espírito Santo there is a population of German settlers that has lived for several generations while remaining immune to miscegenation' (Giemsa and Nauck 1950a: 451–452). 'In the middle of the forests of Brazil, in the remote mountainous valleys in the state of Espírito Santo, there live to this day the descendants of German immigrants who have remained German in their language and their beliefs, their blood and their customs', wrote Nauck in one of the publications about the expedition (1937: 18).

The belief that the colonies in Espírito Santo had not mixed with the 'Luso-Brazilian' population – as the Germans generically referred to the other population segments in Brazil – appears to have been based on the observation of phenotypic characteristics and the conservation of cultural traits rather than on detailed genetic and anthropological research.

Nauck and Giemsa also gave the reason that the region had already been visited by Ernst Wagemann, who had produced a dense report on the living conditions of the settlers exactly twenty years earlier, with the same purpose of studying the question of acclimatization. As already mentioned, the researchers from Hamburg presented their observations as an update to Wagemann's report, which contained minute observations on the social, economic and cultural organization of the community and statistical studies of the population (see Wagemann 1915). Wagemann had justified his interest in Espírito Santo differently, saying that it was one of the largest German settlements in the tropics (18,000 individuals at the time), and it had existed for about seventy years, allowing him to observe three generations of colonists including the first immigrants. In addition, the region occupied by immigrants was located well above sea level (400–800 metres) and in a border area of the tropics (ibid.).

The state of Espírito Santo is a narrow coastal strip of land surrounded by the states of Rio de Janeiro, Minas Gerais and Bahia. Wagemann's describes the territory in 1912 as follows:

> It is a predominantly mountainous region, crossed by numerous waterways, with thick forests. To the north it descends, with some regularity, towards the coast; and to the south there are some hills reaching heights of 1,400 metres. Only the coast is flat, but that region is small. However, above 20°, the plain expands, penetrating further inland, especially in the lower reaches of the Rio Doce, where it spreads, covered with lakes and marshes (ibid.).

The flow of German immigrants to Espírito Santo began in 1847, when the colony of Santa Isabel was founded. Religious differences caused the Protestants to move away and found the village called Campinho. The Pomeranians arrived from 1859, leaving the region that then belonged to Prussia. They migrated in large numbers, came to predominate over the other groups living in the region, and imprinted the mark of their culture there. They concentrated in the Santa Leopoldina colony, founded at the initiative of the Brazilian Emperor Dom Pedro II in 1857. In the following years, new waves of immigrants, and the growth of the initial villages, caused the population to spread towards the mountainous regions located south of the Rio Doce, crisscrossed by its tributaries.

Most Germans settled in the rugged highlands, mostly located in the southern part of the state. The last wave of immigrants arrived in the 1880s, so population growth was due almost entirely to the high fertility rates recorded with surprise by Ernst Wagemann in 1912 (ibid.). As the nineteenth century became the twentieth, the German-born settlers began to spread towards the lowlands to the north and towards the coast. The Pomeranians from Santa Leopoldina occupied lands to the west and soon crossed the Rio Doce heading north, toward the states of Bahia and Minas Gerais (ibid.).[5]

When Nauck and Giemsa visited the region in 1936, this process was in full swing. The advance of the settlers through the lowlands north of the Rio Doce was related to their agricultural methods, as will be discussed later. Communication difficulties between the colonies, dispersion, isolation and the fact that most colonists had not mastered Portuguese appear to have contributed to the timid participation of the German communities in Espírito Santo in the social and political dynamics of the region and in Brazil in general.

I will now examine the role of the Hamburg Institute for Maritime and Tropical Diseases in the trip to Espírito Santo, the interests that led them to make it and the support network that ensured that it took place. I argue that the 1936 expedition to Espírito Santo was part of a strategy pursued by Peter Mühlens, director of the Institute for Maritime and Tropical Diseases,[6] to take an active part in the colonial movement and ensure a leading position for his institution in the organization of public health services in the colonies that Germany was bound to regain through Hitler's foreign policy efforts. Mühlens' strategy regarding his career and the institutional agenda is well described in Stefan Wulf's work on the Hamburg Institute between the Weimar and National Socialist periods (1994).

The Hamburg Institute for Maritime and Tropical Diseasesand the trip to Espírito Santo

The justifications presented by Nauck and Giemsa for the selection of Espírito Santo as a destination do not mention the fact that the trip had been planned in 1934 by the geographer and National Socialist ideologue Oskar Schmieder.[7] Schmieder planned an expedition to the German colonies in Espírito Santo with a team that would include specialists in different fields, with the tropical medicine expert being charged with studying the incidence of diseases in the region (Nauck and Giemsa n.d.). The overall aim was to study the settlers' way of life in order to establish whether it was possible for Germans to colonize the tropics. Allotted a budget of 60,000 Marks, its members would also include a physiologist, a geographer specialized in agricultural studies, a cartographer, a social psychologist, an anthropologist and a tropical public health specialist. Ernst Nauck, a researcher at the Hamburg Institute for Tropical and Maritime Diseases since 1923, was recommended for this last position (PAAA n.d.).[8] At the time, Nauck was writing his thesis for the University of Hamburg on 'Racial Pathology and Geographical Medicine' (Wulf 2010b).

In July 1934, the director of the Institute for Tropical and Maritime Diseases, Peter Mühlens, wrote to Brazilian researcher Henrique da Rocha Lima, who had worked for eighteen years at the Hamburg Institute (Mühlens 1934). Rocha Lima had returned to Brazil in 1927 and in 1933 became director of the Biological Institute of São Paulo, continuing as one of the most active supporters of scientific exchange between Brazil and Germany (Silva 2011: 632–678). In his letter, Mühlens declared his satisfaction with the *Tropeninstitut*'s participation in Schmieder's expedition and asked for Rocha Lima's collaboration on the venture.

The interest in participating in the expedition to Espírito Santo was related to Mühlens' commitment to legitimizing himself in the eyes of the National Socialist leaders and ensuring the outstanding participation of the Hamburg Institute in colonial policy. Mühlens took over as director of the Institute for Maritime and Tropical Diseases in 1934, after a heated leadership contest with public health specialist Ernst Rodenwaldt. The latter was preferred by the National Socialists due to his writings on racial hygiene – widely circulated in the 1920s and 1930s – and his fervent anti-Semitism (Wulf 1994: 83–84; Hansen 2007).[9] On the contrary, the National Socialists saw Mühlens as not sufficiently versed in their racial hygiene concepts and population policies, considered him an old-style conservative and disapproved of his relations with Jewish banker Max Warburg.[10] Mühlens' advantage was his proximity to the Institute's founder, Bernhard Nocht, who intervened with Hamburg's political leaders (Wulf 1994: 84).

For Mühlens and many of his colleagues, regaining the colonies lost under the terms of the Versailles Treaty was a matter of justice. When he took over at the Institute, he made efforts to show the country's new leaders that he shared their political and ideological views and intended to align the institution – and the field of tropical medicine as a whole – with them. He managed to win over his opponents and turn tropical medicine into one of the fields of medicine that was most integrated into the National Socialist machinery, in tune not only with colonial revisionism but also with the 'expansionism founded on the mysticism of the Aryan race' (Eckart 1997: 516). For Mühlens and his peers, Hitler's regime was a perfect opportunity to win back prestige and legitimacy for the field of tropical medicine, established at the turn of the twentieth century in close synergy with European colonialism.[11] The issue of the colonies was initially a bargaining chip in Hitler's diplomatic efforts, especially with the British, and was important in winning the support of those who harked back to the old colonial days, like Mühlens. He started to work closely with the Office of Colonial Policy, created in 1934 and linked to the Foreign Ministry and the Colonial League. In the same year, Mühlens gained a seat on the Reich's Colonial Board. He ensured that the issue of the colonies was a priority in the *Tropeninstitut*'s scientific agenda, making it a factor that determined what direction research should take, what the library should stock, and what specimens should be collected (Wulf 1994: 88f.). A series of trips to Africa was carried out to check local health conditions and investigate the possibility of establishing branches of the Hamburg Institute for Maritime and Tropical Diseases. The territories formerly belonging to Germany were the main destination of the Institute's researchers. Members of the Institute also visited Latin America, not only for reasons relating to the movement to return Germany's colonies to her, but also due to foreign cultural policy interests (ibid.: 103–109).

Although Schmieder's project was evaluated very positively (PAAA n.d.), funding was not approved because of budget restrictions (PAAA 1934). Mühlens then contacted the Hamburg and Berlin authorities in search of backing. The Hamburg Science Foundation granted him 6,000 marks, and the Ministry of Science, Education and Culture pledged a further 1,500 marks. The disparity between these amounts and the original budget indicate that the expedition had been scaled down: rather than the broad-based research originally envisaged by Schmieder, it would specifically address the occurrence of tropical diseases and the impact of the local environment on the people of German descent. Along with Nauck, the only other people to take part in the expedition were the chemist from the *Tropeninstitut*, Gustav Giemsa, and his wife, who actually sailed to Brazil earlier, in November 1935. Rocha Lima had hired Giemsa to organize the chemistry department of the Institute he ran in São Paulo (Ribeiro 1997: 86).[12]

In addition to the financing from the Hamburg Science Foundation and of the Ministry of Science, Education and Culture, Nauck and Giemsa's travel was supported by the network of consulates and embassies, by the National Socialist Party Foreign Relations Organization, and by the Bayer Company. The trip took place between March and May 1936, during a period of intense political, economic and cultural relationships between Germany and Brazil. The soil scientist Paul Vageler (1882–1963) began working at the Agronomy Institute of Campinas, a traditional agricultural research institution in the state of São Paulo, in 1933. The zoologist and geographer Otto Schulz-Kampfhenkel arrived in Brazil in 1935 with Gerd Kahle and Gerhard Krause. They explored the Jari river basin of the Amazon region, on the border with French Guiana, until 1938. The zoologist, ethnographer and anthropologist Hans Krieg visited the states of Paraná, Mato Grosso and São Paulo between 1937 and 1938. In 1936, the São Paulo School of Medicine hired the German pathologist Walter Büngeler, who worked there until 1942 and was an active supporter of the medical exchanges between Germany and Brazil. He actively contributed to the intense participation of Brazilian professors and students of medicine in excursions promoted by the Germano-Iberoamerican Medical Academy to visit medical institutions and to attend specializations courses.[13]

Nauck and Giemsa's trip to Espírito Santo coincided with a period in which the colonial movement was gaining ground in Germany, inspired, amongst other things, by Hitler's speech during the occupation of the Rhineland in March 1936, in which he clearly expressed his intention to win back Germany's lost colonies (Linne 2008: 28). In 1936 the Reich Colonial League was founded, combining all of the entities involved in the colonial movement. Mühlens had a seat on its board. The new association became the official channel for managing all issues relating to colonial revisionism. Thenceforth, there was a flurry of propaganda activity and the first steps were taken to prepare staff for administration of the colonies (ibid.: 46–48). This newfound enthusiasm helped to reignite discussions about the influence of the tropical environment on Europeans as plans were drawn up to establish colonies in the tropics. Let us now follow Giemsa and Nauck's steps up hill and down dale in the state of Espírito Santo, following their reports and observations.

Giemsa and Nauck in Espírito Santo

Ernst Nauck disembarked in Santos, Brazil, in March 1936. He joined Giemsa in São Paulo, where the leader of the National Socialist Party in Brazil, Hans Henning von Cossel, received them. In Rio de Janeiro they visited the German embassy, the Oswaldo Cruz Institute, and the German-Brazilian Institute for High Culture (Instituto Teuto-Brasileiro de Alta Cultura).[14] From Rio de Janeiro, they set off for Vitória, the capital of Espírito Santo, where they visited the German consul, and then travelled to the region that had been colonized by Germans (Giemsa and Nauck 1950a: 452).

The difficulty in accessing the settlements dotted around the region meant that they had to restrict their visits to places that they considered representative of German colonization. They chose a few of the older communities in the higher regions, as well as some more recent settlements at lower altitudes. The area then occupied by the German colonies spread from the highest valleys in the south to the warmer lowlands in the north (ibid.: 455–457).

Continual agricultural expansion had the effect of spreading out the colonization of the region in such a way that there were no densely populated areas. Although this was due to the narrowness of the valleys in the highlands, even at lower altitudes there were no villages. The only thing that gave them a sense of community was their organization into protestant or catholic communities, wrote the researchers (ibid.: 457). The isolation of the settlers and the shortage of funds meant Giemsa and Nauck were unable to study a great number of individuals. They therefore picked around forty families for their study (ibid.: 453f.).

Based on numbers provided by Wagemann and statistical calculations based on church records, Giemsa and Nauck estimated the German-Brazilian population in Espírito Santo to be around 30,000 (Giemsa and Nauck 1950b: 566–568). Based on statistical deductions and information supplied by other authors, they calculated the natural growth rate of their target population group at around 4 per cent, which was very good even when compared with the growth rate in Germany (ibid.: 574). The low mortality rate and large number of children were indicative of admirable population growth, even by the standards of 'civilized countries', as written by Giemsa referring to European countries (1937a: 209).

Giemsa and Nauck recorded a broad range of aspects of the Brazilian-Germans' way of life: their housing, clothing, diet, health, customs, religious life, economy, etc. The colonists' main source of income was coffee farming, which was sufficient for them to buy the other things they needed (1950a: 459–460). However, Giemsa and Nauck were critical of their farming practices, which they claimed consisted of no more than 'cleaning' (weeding and sweeping) the coffee plantations. They were also critical of the practice of expanding their plantations into virgin territory, cutting down native forest and abandoning land as soon as it ceased to be productive rather than fertilizing the soil. They warned that such practices could quickly lead to the complete devastation of the native plant cover, and that a 'more efficient farming system' (ibid.: 460) should be introduced. 'We are facing major problems, because future economic activity will be crucial for the fate of these people of German origin' (ibid.: 461), wrote the scientists. This problem was not unrelated to the race issue, because the constant quest for uncultivated lands 'stimulates nomadic habits, the separation of families, mixing with other ethnic groups, and weakening of racial inheritance' (ibid.).

Expansion of the colonized area towards the lowlands was also a cause for concern, since the climate was less favourable there (hotter and more humid) and there was a greater chance of cross-breeding with non-German people. This migratory process towards the lowlands near the Doce River was already underway. More effective exploitation of the lands already colonized would help to increase the population density, foster greater family union and improve education and society, thereby 'strengthening ethnic inheritance and raising the cultural level' (ibid.), they held.

The centrality of religious structures in the organization of life in the colonies was highlighted by Giemsa and Nauck (1950c: 655–657), who recognized that they owed the good reception they had received in the communities they visited to the help of the priests and ministers, who also provided 'helpful information' (ibid.: 656). They added that the religious communities regulated almost all matters in the community, operating as a kind of 'rural self-government' (ibid.). The Lutheran leaders had for many years engaged in 'dedicated work of valuable cultural nature' (ibid.). The sense of belonging to the German fatherland was kept alive culturally, the scientists found, thanks to the work of these ministers, whose influence gave them an 'in-depth view of the life and nature of the members of the community', exerting 'great influence' (Giemsa 1937a: 205). 'The religious organizations safeguard German customs' (Giemsa and Nauck 1950c: 656).

The religious leaders were also responsible for organizing the educational system. In the catholic communities, Brazilians and Germans attended lessons given primarily in Portuguese and attended the same religious festivities and services, which was not the case in the protestant communities. In addition to cross-cultural contact in the schools and the church, there were some Brazilian doctors, traditional healers and others who treated the members of the immigrant community. They were also required to do military service with the Brazilian army. Only recently, wrote Giemsa and Nauck (ibid.: 662), had the members of the German community realized the advantages of learning the local language, which in many places was taught alongside German (ibid.: 661f.). The ministers were the only link the settlers had with Germany, and efforts were made to raise the level of education, which was generally considered very low. The adults could only read and write German with difficulty, their general knowledge was limited, and although they could read the book of prayer and religious pamphlets, they did not read well enough to subscribe to a German-language newspaper (ibid.: 657).

According to Giemsa and Nauck (ibid.: 662f.) this limited intellectual development was reflected in the cultural inertia of the settlers, who, in the absence of any external stimulus, maintained certain characteristics that set them apart 'as an ethnic group of German origin' (ibid.: 663). While they maintained certain negative 'Pomeranian character traits' – 'stubbornness and retarded mental development' – they also had some positive qualities, like 'constancy, perseverance, pride, a sense of the practical things in their simple lives' (ibid.). Their mistrust of and isolation from Brazilians had, according to the researchers, helped them to conserve their traditional German customs, as seen in weddings, funerals and festivities, as well as the sense of order and attachment to the language. 'Despite all the mistakes, all the mental limitations, and the simple spirit of the settlers, there has remained, through uses and customs, a German cultural heritage whose nature has remained intact in the midst of a foreign world' (ibid.: 664), concluded Giemsa and Nauck.

Of particular note were the observations on the German settlers' health, which was generally considered relatively good. They did not suffer from serious epidemics of diseases like yellow fever or malaria, or venereal diseases or tuberculosis, and the scientists considered the low prevalence of malaria to be of 'decisive importance for the colonization and multiplication of the German community in Espírito Santo' (Giemsa and Nauck 1950b: 562). They mirrored Bernhard Nocht's view that, together with ancylostomiasis, malaria was the biggest hurdle to European settlement in the tropics (1910: 281). Systematic examinations of school-age children did, however, reveal a high prevalence of worm infections, especially hookworm. They attributed this to the consumption of dirty water, absence of sewage treatment, poor diet, and the habit of walking barefoot and bathing in the rivers. Their understanding of hygiene was generally quite basic, said Giemsa and Nauck (1950b: 560). Healthcare was very limited, because the settlements were so hard to reach that regular visits by doctors were unfeasible. Generally speaking, though, the low mortality rate indicated that their state of health was good.

Giemsa and Nauck correlated the families' economic level with the degree of hookworm infection (ibid.). The mental retardation seen in some children and their short stature and lack of vitality were due to worms, which silently sucked their life force and stamina, and not to degeneration of the race: 'We cannot confirm the widespread view that there are many degenerate families amongst the Germans. Some families of limited economic means or given to drinking are said to be degenerate, where diseases, especially ancylostomiasis, are far more serious' (ibid.: 564f.), the scientists stated. The alcoholism they observed in all of the colonies did not show any sign of being inherited. Indeed, no hereditary diseases were observed amongst the families studied (ibid.: 565).

In order to analyse the effects of the tropical environment on the settlers' constitutional heritage, Giemsa and Nauck conducted anthropometric measurements of their heads and bodies and recorded their physical characteristics, such as eye colour, hair type and colour, etc. As they were unable to study all of the individuals, they were forced to select families they regarded as representative of the population group. The distances between the communities, the short length of their stay, and the difficulty in accessing some areas also restricted the scope of the study. They tried to reconstruct family trees back to the generation that had emigrated for some of the forty families in the studies. They also had the settlers answer questionnaires about their origins, ancestors, number of children, health, property, cost of living, etc. and observed them as they went about their daily business. 'Genealogical surveys and surveys of kinship in Espírito Santo could provide invaluable information, allowing comparison of groups with similar backgrounds from the same area of origin settled in different environments, i.e. the same kin and families in Germany and in Brazil' (Giemsa and Nauck 1950c: 653–654). The anthropometric measurements and genealogical studies could serve as a starting point for this enterprise, but could not be taken as 'general conclusions' because this would require researching a greater number of generations (ibid.: 654).

More systematic measurements were made of school-age children. The Hamburg researchers compared their results with measurements of children from Germany, and concluded that the children in Espírito Santo were on average taller than children from northern Germany, but they weighed less and were thinner (ibid.: 655). They suggested that in Espírito Santo, as in other tropical regions, growth of German immigrants compared with Germans in Europe was accelerated until six years of age, when it started to slow down. In other words, as of age seven, the height and weight of the children of German descent living in Espírito Santo started to lag behind that of the children of the same age living in Pomerania. However, according to Giemsa and Nauck, this slowed growth rate was phenotypical in nature – caused by external factors – and not genotypical (genetic):

> It is quite impossible that this is a case of modification of physical features, of a genotypical nature, after so few generations. It is, rather, a modification of the phenotype, to which, alongside the influences of the climate and nutrition, probably the most significant contributory factor is the worm infection in the children in general (ibid.).

No notable difference was observed between the adults in the colonies and adults in rural Germany. In both cases they tended to be slender in build and rarely carried any excess fat. The researchers conjectured that this may be an indication of adaptation to the Brazilian environment, brought about by the climate, the hard work in the fields and the adoption of local eating habits (ibid.).

Their good natural growth rate, reasonable health and physical vitality and the absence of signs of degeneration indicated that the German immigrants had adapted to local conditions. Strong, sturdy families were a sign of the 'qualities of the hereditary inheritance they transplanted to Brazil' (ibid.: 653). The genealogical research performed for one of the families, the Seibels, indicated that they were well-suited to the task of colonization: 'The great intellectual vivacity inherited from the Rhineland side is supplemented by a beneficial mix of perseverance and tenacity, characteristic of Pomeranians' (ibid.: 654f.), concluded Giemsa and Nauck. But what were the broader implications of these peoples' apparently successful adaptation in the debate about acclimatization, newly galvanized by the colonial movement and conceptions of racial hygiene?

'Racial acclimatization' and the Germans in Espírito Santo

Together with the observations of the economic, social and cultural conditions of the German populations in Espírito Santo, Nauck and Giemsa argued that it would be possible to determine to what extent climatic and public health factors influenced the 'maintenance and development of the race transplanted to the tropics' (Giemsa and Nauck 1950a: 451).

Illustrative of the strong connection between scientific debate and colonialism as far back as the eighteenth century (Livingstone 1987),[15] the issue of acclimatization reared its head again in Germany during the National Socialist regime, galvanizing tropical medicine specialists, geographers and anthropologists and providing a common basis for Germany's conception of their colonial policy.

The social historiography of science and medicine has noted how studies of acclimatization illustrate the way in which political and social interests and cultural ideas gave form to scientific theories, operating in close synergy with colonialist ambitions and efforts. According to David Livingstone, the question of human acclimatization has been so widely discussed among scientists that it 'intersects ideologically with a range of important topics in the history of science more generally' (1987: 359). According to this author, 'the implications of acclimatization theory for the European colonization of the tropics were openly acknowledged' (ibid.: 365).

In the late eighteenth century and throughout the nineteenth century the debate about the acclimatization of races was closely tied to the very feasibility of colonization efforts (Livingstone 1987; Harrison 1999; Anderson 1996a, 1996b, 2006). In the mid-1800s, racial theories derived from medical geography and conceptions of health and disease tended to nurture a pessimistic view of the process. When transposed from temperate zones to tropical climes, the race would degenerate physically, psychically and morally, leading people settled there to regress to a lower level of civilization (Anderson 2006). By the turn of the twentieth century, the idea that diseases were caused by infection by specific germs or, in 'tropical diseases', complex parasites that lived their life cycles in other beings, helped to turn the spotlight away from the direct influence of climatic and environmental factors. Even so, susceptibility or resistance to diseases was seen as being particular to each racial constitution (Anderson 1996b). But despite this, the idea that the systemic degeneration of the organism could be triggered by the climate was not completely abandoned and acclimatization continued to spark controversies in the early 1900s.

Herein lies the ambivalence which resonates in Nauck and Giemsa's discourse: while tropical medicine was becoming institutionalized in Europe, based on the concepts and practices of bacteriology, parasitology and modern public health, at the same time it remained profoundly anxious with the proximity of 'native' populations and preoccupied with racial questions. As Warwick Anderson (2006) notes, the idea of 'germ reservoirs' served to stigmatize and segregate the local population, while theories founded in physical anthropology and even medicine continued to emphasize factors linked to heredity. According to Anderson, 'Heredity had once implied a temperament, a characteristic mode of response to discomforting circumstances, a special predisposition to disease, whatever the origin; now this modulating role was redefined in narrower terms of immunity or susceptibility to specific microbes' (ibid.: 97). While tropical medicine optimistically hailed the possibility of controlling the main threat to European settlement in the colonies – tropical diseases – conjectures about the acclimatization of white races in the tropics in general tended to cast it as impracticable.

Historian Wolfgang Eckart states that, from the outset, tropical medicine in Germany took root with its sights set on the issue of racial acclimatization, and was 'firmly located within the direct influence of the German colonial movement' (1997). The issue of acclimatization was one of the reasons why Ernst Wagemann had visited the region colonized by Germans in Espírito Santo in 1912. He argued that doctors were generally inclined to defend a positive view, while anthropologists and geographers held that factors such as radiation, heat and humidity made it impossible for Europeans to become acclimatized (1915). Although this diagnosis might seem oversimplified (tropical medicine specialists were also reluctant to accept the optimistic viewpoint), it is indicative of the uncertainty and controversy that surrounded the subject. In 1912, the International Colonial Institute declared: 'Our knowledge is still limited. What we know without a doubt is that the white race has to struggle to adapt to the tropics, but we do not know whether these difficulties can be overcome or are relative' (97). Two years earlier, acclimatization had been debated at the Third Colonial Congress in Berlin. In it, one of the founders of German tropical medicine, Bernhard Nocht, argued in favour of the more optimistic conception of the process (1910). He distinguished individual acclimatization from racial acclimatization, defining the latter as 'the healthy adaptation not just of the first immigrants, but also of their descendants, so that they are reproduced from generation to generation without mixing with native blood' (ibid.: 279).

Nocht believed that acclimatization of white people was only impossible in areas where tropical diseases were endemic, especially malaria and hookworm infection. It was these, rather than climatic factors, that were the main hurdles to colonization. Even in coastal African countries, seen as the most challenging for European settlement, generations of immigrants could be maintained for decade after decade provided they did not mix with the locals and the region was free of tropical diseases. Also, there should be 'very favourable' (ibid.: 285) economic circumstances. The inferiority of mixed-race people, argued Nocht, was caused just as much by their debilitated inherited constitution as by factors such as education and social influences. In a favourable economic environment and with good education there would be no reason for Germans to fear losing their 'superior qualities' (ibid.: 287f.) in the tropics. Nocht was in favour of 'gradual' acclimatization, as seen amongst Europeans born in the tropics, who interspersed their time in the colonies with regular stays in Europe.

'It is only to be expected that the management of our tropical colonies should mostly be achieved by this means' (ibid.: 289), he maintained.

Nocht's arguments seem to have inspired Giemsa and Nauck's approach to the issue of acclimatization. They alluded to Nocht's distinction between individual acclimatization and racial acclimatization, and added that the acclimatization process should cover not just the adaptation of the body and health, but also the 'transformation of cultural life', which should include 'psychological and cultural adaptation' (Giemsa and Nauck 1950c: 665). They held that acclimatization could not be the only measure of colonization, since the results of colonization depended 'not just on climatic conditions, but also on the physical and spiritual capacity of the immigrant; economic circumstances and methods, the potential for transporting and trading goods, and last but not least, education and schooling' (ibid.). Giemsa and Nauck believed that Espírito Santo illuminated the ways in which groups of German descent could adapt to the tropics. In racial and cultural terms, the settlers had maintained the distinctive traits of the German people. Their 'good constitution' and high natural growth rate suggested that they presented signs of acclimatization:

> Despite the occurrence of certain diseases, especially worm infections, and some climatic damage arising from the extension of the colonized areas to the low-lying lands, there are indications of acclimatization, such that the continued existence of the population seems assured, at least from a health perspective (Giemsa 1937a: 214).

The Espírito Santo settlements were illustrative of the fact that the region was almost the only one in which the Germans had not mixed with the Brazilian people and 'gone native', thereby demonstrating the factors that would make adaptation to a tropical environment successful (Giemsa and Nauck 1950c: 665). The observations made in Brazil cast light on one of the most debated features of acclimatization: whether white people would maintain their capacity to do heavy manual labour on an on-going basis in the tropics. Giemsa and Nauck recorded that the German settlers in Espírito Santo worked hard from childhood to an advanced age, doing heavy manual work such as chopping down trees, clearing undergrowth and doing woodwork for around eight hours a day and even up to twelve hours a day during the coffee harvest (ibid.: 666). The intense physical strain, they found, was offset by sound sleep, good food and the 'simplicity and regularity of their lives' (ibid.). If they had 'degenerated', this would be 'apparent', and would not be caused by the climate or by alterations to their genetic constitution, but by phenotypical modifications caused by diseases or other environmental factors. Their cultural 'backwardness' could be explained by their geographical and social isolation and poor schooling (ibid.: 667). The Espírito Santo case demonstrated eloquently that social, economic and cultural processes were as important to the success of acclimatization and colonization as maintenance of genetic and hereditary traits:

> If (despite the lack of intellectual orientation, despite limited education, unfavourable economic circumstances, and lack of medical care and hygiene)

> the people have withstood the climate, have done heavy manual labour as well as could be hoped, have settled in new areas, remaining German in their nature, ways, language and beliefs, revealing over several generations the capacity to work and reproduce, we may consider that the assumptions that in our view compose the concept of successful acclimatization have been fulfilled (ibid.).

Maintaining racial purity seems to have been an absolute precondition for any discussion of acclimatization to the tropics. While it was already present in Nocht's discourse in 1910, it gained extra impetus in the intellectual climate of the Third Reich. Giemsa and Nauck's conclusions were based on Nocht's arguments, but also acknowledged the more radical thinking that gained ground in tropical medicine during the National Socialist years, whose main exponent was Ernst Rodenwaldt. According to Wolfgang Eckart, Rodenwaldt was the 'leading protagonist in anthropology and racial hygiene in Germany in the 1930s and 40s', and 'combined the personality of a military man and intellectual, fanatically defending racial segregation' (1997: 521). In the context of theoretical debates, he spoke for all who stressed the importance of factors affecting the individual constitution (genetics and race) in the disease incidence, and criticized those that focused exclusively on fighting parasites and vectors (Hansen 2007: 43–54). The focus of his criticism was the staff of the Hamburg's Institute.

Two years after Giemsa and Nauck visited Espírito Santo, Rodenwaldt addressed the issue of acclimatization in his manual on 'Tropical Hygiene' (*Tropenhygiene*). Both researchers had presented their observations in 1936 at the annual meeting of the German Society of Tropical Medicine, where acclimatization was the main topic on the agenda. In 1937, they also published them in the official journal of the *Tropeninstitut*, and published them individually and jointly elsewhere (Giemsa 1937a, 1937b; Giemsa and Nauck 1937, 1939; Nauck 1937, 1938a, 1938b, 1939). Without directly alluding to their work, Rodenwaldt warned that individual success cases of acclimatization could not be taken as a general rule, since the tropics were not one, but a mass of many different microenvironments with their own very particular characteristics (1938: 134–135). He argued that people who settled at higher altitudes tended to fare better because the climate was more similar to that of temperate zones. 'Admirably high' (ibid.: 132) birth rates, such as those seen in Espírito Santo, could equally not be seen as guarantees of successful settlement by Europeans in the tropics. Furthermore, if 'adaptation' was really what had happened, it still meant regressing to a culturally inferior level, since the German people who migrated to the tropics became culturally stagnated, more often than not turning into mere 'caricatures of a living European community' (ibid.: 133). 'Nowhere can one see amongst these people living in the tropics any development or progress to be compared to what is seen in the fatherland' (ibid.), he argued. And he continued:

> If, in their new tropical homeland, pure-blood Europeans, even under favourable economic circumstances, refuse, or even do not consider

> building latrines and for this reason suffer from hookworm like the country's natives; if their garb and diet remains unchanged from the time they emigrated, this all indicates the absence of any forward thinking in racial terms, the diminishment of all vitality and all desire to progress [...] In this sense, degeneration denotes the process by which settlers descended from a gifted, vibrant European people cannot raise themselves culturally, economically or politically in a colonial country and, rather than becoming models and repositories of culture, remain as mere objects of the administration of a foreign people, to wit, mestizo (ibid.).

It is reasonable to suppose that Rodenwaldt was here referring to Giemsa and Nauck's work. From a racial hygiene and hereditary research perspective, it was very unlikely that any changes would have occurred in the genetic and racial makeup of people settled in the tropics, Rodenwaldt believed, but the cultural stagnation and continued threat of tropical diseases did not allow for an optimistic appraisal of acclimatization.

In the full report on their work published in 1939, Giemsa and Nauck reflect on some of Rodenwaldt's comments, especially the mistake of taking the tropics as one homogeneous environment (1950c: 664f.). They agreed that isolated cases of successful acclimatization had no bearing on the broader discussions about acclimatization of Europeans to the tropics. Treading a fine line between the conceptions of racial hygiene and representing the Espírito Santo settlers' way of life as comprehensively as possible, Nauck and Giemsa's discourse is tainted by ambivalence throughout, constantly making reference to the lack of clarity and grey area that existed between the dimensions of 'race' and 'culture', even in a context so strongly marked by racial ideas as National Socialist Germany.

Acclimatization and the Third Reich's Colonial Project

Despite Giemsa and Nauck's wariness about extrapolating their analyses, their statements suggest that the tropics could be colonized by people of German descent. But they quickly warned that they did not intend to encourage mass migration to tropical regions. 'The German people cannot give up their valuable elements' (Giemsa and Nauck 1937: 21), they stated. Germany could not repeat the mistakes of the past, when millions of people had left the country only to be incorporated into their new lands through cross-breeding (ibid.: 21). Rodenwaldt warned of the risks involved in German colonization in the tropics: only a few families must take on the tasks of colonial administration, and must be strictly answerable to the racial laws (Nuremberg Laws) in place in Germany since 1935 (1938: 134). Giemsa and Nauck held that the colonial lands Germany strove for should serve initially for the exploitation of raw materials, and mass colonization by 'valuable Germans' should not be encouraged (1950c: 668); a recommendation confirmed by Mühlens in the foreword to a report published in 1939: 'It is true that the colonial space claimed by us as a right should rather serve for the exploitation of raw materials than for colonization. Mass colonization of the tropics by valuable ethnic Germans is to be curbed for reasons of national political order'. The scientists towed the line set by government and National Socialist leaders involved in the colonial movement (Linne 2008: 31, 41). Only a few Germans endowed with a good 'hereditary constitution' (Eckart 1997: 518) and complying with the strict racial hygiene and segregation laws would dominate the native workforce and coordinate the supplies of raw materials. Giemsa and Nauck even suggested that a few families in Espírito Santo might be suitable for such a task in the colonies (1937: 21).

The full text of Giemsa and Nauck's report was only published in 1939 in the University of Hamburg's publication on foreign cultural studies (*Abhandlungen aus dem Gebiet der Auslandskunde*). In the preface, they state:

> Many circumstances have delayed the printing of our report, which we publish only now. In the space of time that has elapsed between our voyage and the appearance of this book, nothing we deem to be significant has changed in the demographic and biological conditions. The political circumstances and future potential for development that may be influenced by them have altered in recent years and been taken into account insofar as they pertain to our research (1950a: 452).

In 1939, preparations for future colonial administration were given a new boost on the orders of Hitler himself (Linne 2008: 48). The Foreign Ministry was in charge of presenting and negotiating the return of the colonies on the diplomatic level. The Institute for Maritime and Tropical Diseases in Hamburg was responsible for training the doctors who would work in the colonies. In 1938, Mühlens made a trip to Africa after which he submitted plans for the creation of a branch of the *Tropeninstitut* there. The Reich Colonial Board also undertook to form a committee to train colonial doctors (ibid.). In May 1939, a nine-week training course for doctors commenced. The Institute of Tropical Medicine, headed by Rodenwaldt, an institution linked to the Military Medical Academy, competed with Hamburg over the right to train colonial doctors. With war on the horizon, the return of the colonies was now being pursued by military rather than diplomatic means. The outcome of the conflict, however, put paid to all of Germany's colonial aspirations, though not without first causing the criminal, cruel death of people seen as inferior and 'unfit' to belong to the 'new Germany'; a brutal corollary to the concepts of racial hygiene communicated and disseminated at the time as legitimate science.

Conclusion

This chapter has described Nauck and Giemsa's expedition to Brazil, relating it to the scientific and political context of the Third Reich. The legitimacy and communication of their statements about the acclimatization of Germans in the tropics were closely linked to the rise and fall of colonial ambitions at the time.

During the Third Reich, the acclimatization issue was one of the channels by which German tropical medicine specialists joined the debate on racial hygiene and racial politics, defending the application of their precepts in the formation and administration of the new colonies. It also represented one of the ways in which tropical medicine was integrated into the National Socialist political system; an integration that, again according to Eckart, was highly successful (1997: 516).

Nauck and Giemsa's trip to Espírito Santo was part of Mühlen's strategy to ensure a leading position for his institute at a time when tropical medicine was marked by rifts, rivalries, contentions and contradictions. The appropriation of the ideas of racial hygiene and biological anthropology reveals an effort to build consensus at a time when the colonial movement was winning over more people and political support. Its emphasis on disease and economic and cultural factors lends Nauck and Giemsa's discourse a certain ambiguity and, stripped of the racial lexis and dogma of the Third Reich, provides an interesting picture of the life led by people of German origin in Espírito Santo. This was recognized by one of Brazil's leading historians, Sérgio Buarque de Holanda, who published his masterpiece, *Roots of Brazil*, in the same year that Giemsa and Nauck visited the immigrant communities. According to Holanda (1950), the presence of 'racial doctrines and even nationalistic ideologies' in the German researchers' report could only be of interest to certain Brazilian intellectuals 'out of wild curiosity' (cited in Ackermann 2005: 9). But for him, the reasons that drove them to do their work in Espírito Santo should serve more as motivation to understand them than to condemn them out of hand.

'Race', a category used as a basis for the political, social and ideological organization of National Socialist Germany, would also be the cornerstone of the administrative approach to be adopted in these colonies. Mühlens and the personnel from the Hamburg Institute did not represent the most radical wing of tropical medicine when it came to the issues of racial hygiene. They were aligned with the more 'traditional' school of thought linked to colonial revisionism as it had taken shape since the Treaty of Versailles and throughout the Weimar Republic. According to Wolfgang Eckart (1997: 519), this current was primarily devoted to laying the necessary theoretical and ideological groundwork for the return to Africa and the Pacific, in which the Hamburg Institute would have an active role in resolving all manner of practical issues. Giemsa and Nauck's conclusions allude to this more traditional branch of German tropical medicine identified with the founder of the Hamburg Institute (also known as the *Tropeninstitut*), Bernhard Nocht, and the way in which he and the institute as a whole tended to consider colonial revisionism and hygiene in the tropics.[16]

Indirectly linked to colonial issues as perceived by this nation stripped of its overseas possessions, Giemsa and Nauck portrayed the diversity of an 'other' that was at the same time the 'self'. They sought to ascertain whether and to what extent those settlers with whom they shared a common tongue and physical features had changed because they were living in a different environment. Would the 'otherness', which David Arnold (1996) takes as the heart of the representation of the tropics, be extended to those who went to live there? Determining the boundaries between immutable hereditary features and traits that could be altered by environmental factors was the main concern in the different theories that sought to make sense of the phenomena of heredity and human diversity. Meanwhile, the connection to the German nation by 'blood and soil', while bringing the settlers of German origin onto a collision course with the nationalist policies of the Brazilian state as of 1937, was a way of ensuring that their Germanness stayed intact, provided their distinguishing features remained unsullied.

The main idea, defended in stronger or weaker versions and present in Giemsa and Nauck's observations, was that individuals with German blood should not breed with non-German people. Echoes of the view of the tropics as the 'white man's grave' resonate in studies which, like the one done by Giemsa and Nauck, sought to determine the effects of the tropical environment on the 'German race'. Only a racially pure population would be capable of avoiding the potentially degenerative effects of the sun, heat and humidity in the tropics and maintain the characteristics that legitimized their dominion over 'native' peoples. It would be up to leaders, abiding by the scientific principles and laws of racial hygiene, to enforce racial segregation in the regained colonies.

The colonial project that Nauck and Giemsa were involved in sought to exploit the human and natural resources in the African colonies and establish an order in which the 'white race', particularly the 'German race', would be the 'master race', controlling the whole system designed to ensure the greatness and expansion of 'Greater Germany'. Implicit in Nauck and Giemsa's views on acclimatization was a criticism of miscegenation. Communities of German descent would only attain a minimum of success in adapting to the tropics if they remained racially pure, meaning they would have to avoid inter-racial intercourse.

The 'success' of the German immigrants' acclimatization, as perceived by the scientists, was more due to cultural, social and economic strategies than to any organic modifications that would have changed their biological endowments. While the researchers may have gone to Espírito Santo as part of a National Socialist racial hygiene agenda and adopted the tools of physical anthropology and heredity research, their approach was actually aligned with the fundamental principles of 'tropical medicine': the investigation of pathogens and vectors and the observation of health conditions. It was not that the settlers in Espírito Santo had degenerated through contact with the harsh environmental conditions or been tainted in their hereditary legacy, but rather that they were just unhealthy, pure and simple. They were condemned, Nauck and Giemsa argued, not by having degenerated in the tropics and thereby tainted their racial inheritance, but by the geographical – and consequently cultural and even temporal – distance that kept them apart from the 'revolution' that was shaping a 'new Germany'. Meanwhile, the researchers also believed that behavioural traits such as a lack of leadership, mental retardation and mistrust were derived from hereditary factors. The conception of race expounded by Giemsa and Nauck trod a fine line between a fatalism that portrayed it as a direct corollary of hereditary features and a vision that also embraced social and cultural factors. The keys to the Germans' success in their colonial settlements in the tropics would be a combination of hard work, maintaining racial purity, and keeping the features that marked them as German, as well as implementing the measures propounded by modern public health and tropical medicine. This was therefore the recipe for success for an effective administrative system in the new colonies. Presenting themselves as experts in racial and medical hygiene in the tropics, the Hamburg researchers sought to operate as arbiters of the new colonial order.

André Felipe Cândido da Silva is a Researcher at the Casa de Oswaldo Cruz/ Oswaldo Cruz Foundation, Science Editor of the journal *História, Ciências, Saúde – Manguinhos* and Professor in the Graduate Program in History of Science and Health, also at the Casa de Oswaldo Cruz. He earned a PhD in History of Science and Health from the Casa de Oswaldo Cruz. His areas of research and teaching are the history of medicine and sciences in transnational perspective, with a focus on the German-Brazilian relations in the first half of the twentieth century, and the history of health, agriculture and environment.

Notes

1. On German tropical medicine during the Third Reich see Eckart (1997) and Hansen (2007: 43).
2. On this theme, see Weindling (1989), Schmuhl (2008) and Hutton (2005).
3. For this chapter I consulted a Brazilian translation published in *Boletim Geográfico* 8(88, 89 and 90). This quotation is from Giemsa and Nauck (1950a: 452).
4. For this work I used a translation from German to Portuguese (Wagemann 1948–1949).
5. See also Giemsa and Nauck (1950a) and Roche (1968).
6. The Institute for Maritime and Tropical Diseases was founded in Hamburg in 1900 in response to the 1892 cholera epidemic. The hope was to create a centre where tropical medicine and hygiene in the colonies could be studied, in much the same way as in Britain (see Wulf 1994; Mannweiler 1998; Brahm 2002). Its first director, Bernhard Nocht, was a disciple of Robert Koch. His work during the epidemic had marked him out for praise and he also had experience in dealing with 'tropical' diseases. The institute invested in researching the diseases most prevalent in Germany's overseas possessions, training the doctors and nurses sent there, and conducting health inspections of retail establishments and the Hamburg port. Nocht ran the institute until 1930, when he retired and was replaced by one of the institute's researchers, Friedrich Fülleborn. The end of Fülleborn's short time as director was marked by disputes with the National Socialist over the running of the institute as of 1933, when they came to power. He died of a heart attack in September 1933, and his demise sparked a leadership crisis between Mühlens and Rodenwaldt (see Wulf 1994: 82).
7. In 1930, Schmieder began teaching geography at the University of Kiel, where he was one of the people who promoted the doctrine of 'living space' (*Lebensraum*) – the ideological substrate for the National Socialists' expansionary ambitions. According to this theory, 'racially superior' peoples should spread beyond their national borders, annexing territories and subjugating 'inferior' peoples to make way for the formation of a community founded on racial identity.

8. Russian-born Nauck joined the institute in 1923, after having taken part in the celebrated German Red Cross expedition to combat the starvation and serious Typhus fever epidemic in Russia in 1921–1922. He lived in China from 1924 to 1927 and then in Costa Rica from 1927 to 1929. In both countries he studied and taught tropical diseases.
9. On Rodenwaldt's career, see Jusatz (1965, 1970), Eckart (1998) and Kiminus (2002). Part of his scientific work and career are described in Mannweiler (1998: 131–132) and Hansen (2007).
10. For more on Peter Mühlens's career, see Wulf (1997, 2010a).
11. For more on the institutionalization of 'modern' tropical medicine and its relationship with European colonialism, see, for instance, Worboys (1993) and Arnold (1996). A more recent review of this literature can be found in Neill (2012).
12. Giemsa was best known for having perfected a staining technique widely used for the observation of parasites and cells and for his work in chemotherapy and pharmacology.
13. On the German expeditions to Brazil at that time, see Sá and Silva (2016). Specifically on Otto Schulz-Kampfhenkel and his expedition to Brazilian Amazon, see Flachowsky and Stoecker (2011).
14. Founded in 1930 by some pro-German scientists and intellectuals interested in fostering academic exchange between Brazil and Germany (Silva 2011: 593–595).
15. See also Anderson (1996a, 2006).
16. A more in-depth perspective on this issue can be found in Wulf (1994).

Bibliography

Ackermann, S. 2005. 'Sob a lente alemã: súditos de Hitler pesquisam populações de origem alemã no Estado do Espírito Santo', in *Anais do Simpósio Muitas Faces de uma Guerra*, Florianópolis: UDESC. Retrieved 23 August 2011 from www.cce.udesc.br/cem/simposioudesc/anais/ st2/st2silvia.doc

Anderson, W. 1996a. 'Disease, Race and Empire', *Bulletin of the History of Medicine* 70(1): 62–67.

——— 1996b. 'Immunities of Empire: Race, Disease, and the New Tropical Medicine, 1900-1920', *Bulletin of the History of Medicine* 70(1): 94–118.

——— 2006. *The Cultivation of Whiteness: Science, Health and Racial Destiny in Australia*. Durham: Duke University Press.

Arnold, D. 1996. *Warm Climates and Western Medicine: the Emergence of Tropical Medicine 1500-1900*. Amsterdam and Atlanta: Rodop.

Brahm, F. 2002. *Die Lateinamerika-Beziehungen des Hamburger Tropeninstituts 1900-1945*, MA thesis in History. Hamburg: Universität Hamburg.

Eckart, W.U. 1997. *Medizin und Kolonialimperialismus: Deutschland 1884-1945*. Munich and Vienna: Paderborn.

——— 1998. 'Generalarzt Ernst Rodenwaldt', in G.R. Ueberschär (ed.), *Hitlers militärische Elite. Von den Anfängen des Regimes bis Kriegsbeginn*. Darmstadt: WBG, pp. 210–222.

Ernst, W. 1999. 'Introduction: Historical and Contemporary Perspectives on Race, Science and Medicine', in W. Ernst and B. Harris (eds), *Race, Science and Medicine, 1700-1960*. London and New York: Routledge, pp. 1–28.

Flachowsky, S. and H. Stoecker (eds). 2011. *Vom Amazonas an die Ostfront: Der Expeditionsreisende und Geograph Otto Schulz-Kampfhenkel (1910-1989)*. Cologne, Weimar and Vienna: Böhlau u. a.

Fleck, L. 1979. *The Genesis and Development of a Scientific Fact*. Chicago: Chicago University Press.

Giemsa, G. 1937a. 'Ein deutschstaemmiges, im brasilianischen Staate Espírito Santo bodenständig gewordenes Kolonistenvolk', *Koloniale Rundschau* 28(3): 200–215.

—— 1937b. 'Gesundheitverthältnisse bei den deutsch-stämmigen Kolonisten im brasilianischen Staate Espirito Santo', *Gesundheit und Leben* 2(7): 11–14.

Giemsa, G. and E.G. Nauck. 1937. 'Rasse und Gesundheitserhaltung sowie Siedlungsfragen in den warmen Ländern', *Archiv für Schiffs- und Tropenhygiene* 41(1): 9–21.

—— 1939. *Eine Studienreise nach Espírito Santo – Volksbiologische Untersuchung einer deutschstämmigen Bevölkerung in Mittelbrasilien als Beitrag zum Akklimatisationsproblem*. Hamburg: Hansische Universität [Abhandlungen aus dem Gebiet der Auslandskunde 48].

—— 1950a. 'Uma viagem de Estudos ao Espírito Santo – Pesquisa demo-biológica, realizada, com o fim de contribuir para o estudo do problema da aclimação, numa população de origem alemã, estabelecida no Brasil Oriental', *Boletim Geográfico* 8(88): 451–470.

—— 1950b. 'Uma viagem de Estudos ao Espírito Santo II', *Boletim Geográfico* 8(89): 560–575.

—— 1950c. 'Uma viagem de Estudos ao Espírito Santo III', *Boletim Geográfico* 8(90) 653–701.

Hansen, F. 2007. *1907-2007: Deutsche Tropenmedizinische Gesellschaft – Eine Chronik*. Hamburg: Deutsche Gesellschaft für Tropenmedizin und Internationale Gesundheit e.V.

Harrison, M. 1999. *Climates and Constitutions: Health, Race, Environment and British Imperialism in India 1600–1850*. New York: Oxford University Press.

Holanda, S.B. 1950. *Clima e Raça*. São Paulo: Folha da Manhã.

Hutton, C.M. 2005. *Race and the Third Reich: Linguistics, Racial Anthropology and Genetics in the Dialectic of Volk*. Cambridge and Malden: Polity.

Institut Colonial International. 1912. *Compte rendu de la Session tenue à Bruxelles les 29, 30 et 31 Juillet 1912*. Brussels: Bibliothèque Coloniale Internationale.

Jusatz, H.J. 1965. 'Ernst Rodenwaldt (08.05.1878 – 06.04.1965)', *Zeitschrift für Tropenmedizin und Parasitologie* 16(3): 233–234.

—— 1970. 'Ernst Rodenwaldt (1878–1965) als Begründer der geo-medizinischen Forschung', *Heidelberger Jahbücher* 14: 23–51.

Kiminus, M. 2002. *Ernst Rodenwaldt, Leben und Werk*, unpublished PhD Thesis. Heidelberg: University of Heidelberg.

Linne, K. 2008. *Deutschland jenseits des Äquators? Die NS-Kolonialplanungen für Afrika*. Berlin: Ch. Links Verlag.

Livingstone, D. 1987. 'Human Acclimatization: Perspectives on a Contested Field of Inquiry in Science, Medicine and Geography', *History of Science* 25: 359–394.

Mannweiler, E. 1998. *Geschichte des Instituts für Schiffs- und Tropenkrankheiten in Hamburg, 1900-1945*. Keltern-Weiler: Goecke und Evers.

Mühlens, P. 1934. 'Brief to Henrique da Rocha Lima of the 29.03.1934'. Rocha Lima Papers. São Paulo: Memorial Centre of São Paulo's Biological Institute.

Nauck, E.G. 1937. 'Deutsche Kolonisten in Espirito Santo', *Gesundheit und Leben* 2(9): 18–20.

—— 1938a. 'Akklimatisation und Siedlung in den Tropen', *Die Umschau in Wissenschaft und Technik* 42(42): 960–964.

—— 1938b. 'Die Akklimatisation und ihre Bedeutung für die Siedlung in den Tropen', *Zeitschrift der Gesellschaft für Erdkunde* 314: 81–93.

——— 1939. 'Zur Frage der Akklimatisation', *Deutsche Medizinische Wochenschrift* 65(26): 1046–1049.

Nauck, E. and G. Giemsa. n.d. 'Reisebericht nach Espirito Santo'. Typed. Historical Archives of Bernhard Nocht Institut für Tropenmedizin, Ordner Institutsgeschichte 1935–1936.

Neill, D. 2012. *Networks in Tropical Medicine: Internationalism, Colonialism, and the Rise of a Medical Specialty, 1890–1930*. Stanford: Stanford University Press.

Nocht, B. 1910. 'Der derzeitige Stand der Akklimatisationsfrage', in Deutscher Kolonialkongress (ed.), *Verhandlungen des deutschen Kolonialcongresses 1910, zu Berlin am 6., 7. Und 8. Oktober 1910*. Berlin: D. Reimer, pp. 278–289.

Politisches Archiv des Auswärtigen Amtes (PAAA). n.d. 'Verschiedenes – Blatt 4', *Akte 65822 – Notgemeinschaft der deutschen Wissenschaft*.

——— 1934. 'Brief an den Professor Schmieder vom 09.06.1934', *Akte 65822 – Notgemeinschaft der deutschen Wissenschaft*.

Ribeiro, M.A.R. 1997. *História, Ciência e Tecnologia: 70 anos do Instituto Biológico de São Paulo na defesa da agricultura, 1927-1997*. São Paulo: Instituto Biológico.

Roche, J. 1968. *A Colonização Alemã no Espírito Santo*. São Paulo: Difel.

Rodenwaldt, E. 1938. *Tropenhygiene*. Stuttgart: Ferdinand Enke Verlag.

Sá, M.R. and A.F.C. Silva. 2016. Citizens of the Third Reich in the Tropics: German Scientific Expeditions to Brazil under the Vargas Regime, 1933-40, in, F. Clara and C. Ninhos (eds), *Nazi Germany and Southern Europe, 1933-1945: Science, Culture and Politics*. London: Palgrave MacMillan, pp. 232–255.

Schmuhl, H.W. 2008. *The Kaiser Wilhelm Institute for Anthropology, Human Heredity and Eugenics, 1927-1945: Crossing Boundaries*. Dordrecht: Springer.

Silva, A.F.C. 2011. *A trajetória científica de Henrique da Rocha Lima e as relações Brasil-Alemanha (1901-1956)*. PhD Dissertation. Rio de Janeiro: Programa de Pós-Graduação em História das Ciências e da Saúde, Casa de Oswaldo Cruz, Fundação Oswaldo Cruz.

Wagemann, E. 1915. *Die Deutschen Kolonisten im Brasilianischen Staate Espírito Santo. Schriften des Vereins für Sozialpolitik 147(5)*. Munich and Leipzig: Duncker & Humblot.

——— 1948–1949. 'A Colonização Alemã no Estado do Espírito Santo', *Boletim do Instituto Brasileiro de Geografia e Estatística*, 68, 69 and 70.

Weindling, P. 1989. *Health, Race and German Politics between National Unification and Nazism, 1870-1945*. Cambridge: Cambridge University Press.

Widmer, A. 2012. From Research Encounters to Metropolitan Debates: The Making and Meaning of the Melanesian 'Race' during Demographic Decline, Paideuma: Mitteilungen zur Kulturkunde (58), 69–93.

Worboys. M. 1993. 'Tropical diseases', in W. Bynum and R. Porter (eds), *Companion Encyclopedia of the History of Medicine*. London and New York: Routledge, pp. 512–536.

Wulf, S. 1994. *Das Hamburger Tropeninstitut 1919 bis 1945: auswärtige Kulturpolitik und Kolonialrevisionismus nach Versailles*. Berlin and Hamburg: Reimer.

——— 1997. 'Peter Mühlens. Tropenmediziner und -hygieniker', in *Historische Kommission bei der Bayerischen Akademie der Wissenschaften* (ed.), *Neue Deutsche Biographie* 8. Berlin: Duncker & Humblot, p. 285. Retrieved from http://www.deutsche-biographie.de/ppn101597991.htm.

——— 2010a. 'Peter Mühlens', in F. Kopitzsch and D. Brietzke (eds), *Hamburgische Biographie Personenlexikon* 5. Göttingen: Wallstein Verlag, pp. 268–270.

——— 2010b. 'Ernst Nauck', in F. Kopitzsch and D. Brietzke (eds), *Hamburgische Biographie Personenlexikon* 5. Göttingen: Wallstein Verlag, pp. 272–273.

2
'Ill-suited' Populations in German Nauru
Race, Health and Labour under Company Administration, 1888–1914

Antje Kühnast

In April 1886, the governments of the German and British empires delineated their respective colonial spheres in the Pacific (Great Britain, Parliament, House of Commons, 1886). Due to its geographical location west of the demarcation line, and at the instigation of the Hamburg trades companies the Deutsche Handels- und Plantagen-Gesellschaft and Robertson & Hernsheim, Nauru became part of Germany's colonial dominion. At the time, around a thousand Nauruans and a handful of Europeans inhabited the tiny remote island south of the equator (Sonnenschein 1889: 21). Similar to other societies in the South Pacific archipelago of Micronesia, the Nauruans produced coconut palm oil and copra, supplying the industrializing European markets with important ingredients for cooking oil and soap (Firth 1982: 17; Hardach 2002: 510).[1] Although they had remained relatively isolated from major shipping routes, they had successfully traded with the two competing Hamburg companies for several decades.

In the following year, the German Colonial Office endorsed the merger of these two firms, resulting in the Jaluit-Gesellschaft, and granted it the exclusive rights to acquire land, fish for pearl shell and exploit guano deposits for the production of phosphate fertilizer in the Marshall Island Protectorate. In return, the company paid for the expenses of colonial administration. On 1 October 1888, Nauru was incorporated into the Jaluit-Gesellschaft's entrepreneurial regime – 'a form of government by company' as Stewart Firth has aptly called it (1978: 38; see also Hardach 2002: 514–515).[2]

During the ensuing twenty-six years of German rule (which ended with Nauru's occupation by Australian troops at the beginning of the First World War), the population structure of Nauru and its administration changed. Initially acquired merely for the exploitation of its copra production and its inhabitants as labourers for other German Pacific colonies, Nauru became a highly profitable asset after the discovery of phosphate in 1900 (Firth 1973: 26–27). The utilization of this sought after resource for Australia's and Europe's increasingly industrialized agricultural production (Firth 1978: 36; Denoon 1997: 237) introduced more parties to the colonial power relations on the island: the British Pacific Phosphate Company (PPC) and the Jaluit-Gesellschaft brought in German, British and Australian staff and arranged for indentured labourers from China and the Caroline Islands. Nauru, from then on, experienced a specific model of colonial rule, based on a mix of German imperial and British-German entrepreneurial power relations.

In his analysis of German colonial labour policy in the Pacific, Firth has stated that 'in the early days of the Pacific phosphate industry governments were merely the agents of private companies, providing legality for whatever the companies wished to do' (1978: 37). Accordingly, 'colonial policy was not made by government officials on the spot in the Pacific, whether British or German, but by firms in London, Bremen and Berlin' (ibid.). With regard to Nauru, Firth has demonstrated that 'the aim of the governments in Rabaul and Berlin was to supply the phosphate firms with cheap land, long mining leases, low royalties and a disciplined, underpaid labour force' (ibid.: 51). As a result, between 1906 and 1914, German labour policies in Nauru, under the combined corporate influence of the PPC and the Jaluit-Gesellschaft, authorized the ill treatment of indentured labourers from China and the Caroline Islands and contributed to the high death toll among both the indigenous and indentured populations of Nauru (Firth 1978: 41, 52; Eckart 1988: 94–95).

This high death toll was caused by the mismanagement of recurring epidemics among the Nauruans and the Chinese labourers. As Wolfgang U. Eckart has shown, the medical services in the German Pacific were highly inefficient. This was partly caused by the remoteness of its many islands – especially those subsumed under the Marshall Islands Protectorate – which made comprehensive health care difficult for the few government physicians stationed in the colonies' administrative centres. According to Eckart, a combination of the German government's disinterest in its Pacific colonies and the Jaluit-Gesellschaft's 'uncompromising capitalistic methods of exploitation' (1988: 93) inhibited the establishment of a functioning health care system and thwarted the efforts of generally well-meaning government physicians on the ground (ibid.: 93–97). Thus both the German government's negligence and the mining companies' 'greed' (ibid.: 96) caused the devastating health situation on Nauru.

In extension of Firth's and Eckart's studies, I argue that racial assumptions were deeply ingrained in the binational corporate governance of the indigenous and indentured populations on Nauru. My analysis of the available contemporaneous German written sources about Nauru – anthropological literature, German colonial health reports and the PPC's correspondence[3] – reveals, firstly, that German perceptions of the Nauruans changed according to the context of the different phases of colonization. Namely, German observers initially construed an 'ennobled savage' image of the Nauruans as a (reinvigorated) physically and mentally strong, happy and healthy race. With the beginning of phosphate mining, however, when the Nauruans showed no interest in mining and contracted introduced diseases, they became a weak and unreasonable race doomed to extinction in the eyes of most German observers.

Secondly, I shall show that the inefficient treatment of diseases (such as influenza, beriberi and dysentery) that affected the Nauruans and the Chinese during the first years of phosphate mining was not caused exclusively by the ruthless treatment of humans as cheap labour resource. Essentially, perceptions of cultural, mental and physical difference were intricately embedded in racial assumptions about the incompatibility of certain human groups for particular kinds of work and lifestyles.

Ennobling Nauruans

As Bronwen Douglas has shown, from the beginning of the Europeans' contact with the inhabitants of the South Pacific, their perception and labelling of Oceanic peoples 'was internally racialized, with skin colour and physical organization the key differentiae in the elaboration of region-wide racial taxonomies' (2008: 8). According to Douglas, various 'convoluted schemas' (ibid.: 9) of racial classification were initially devised by European seafarers in Oceania creating a fusion of geographical divisions with physical differentiations (ibid.: 6–14). As Jean Mitchell further illustrates in the

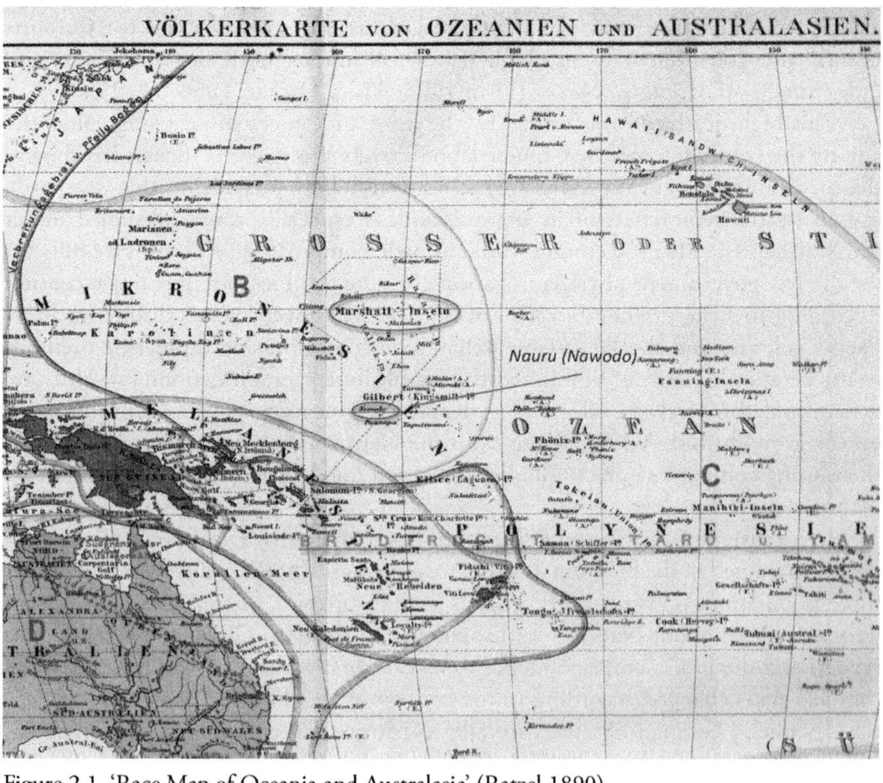

Figure 2.1. 'Race Map of Oceania and Australasia' (Ratzel 1890)

next chapter of this volume, most importantly the French explorer Jules-Sébastien-César Dumont d'Urville broadly distinguished between darker-skinned Melanesians and lighter-skinned Polynesians, Malayans and Micronesians. While the latter were generally deemed physically, mentally, culturally and morally superior to the former, the meaning of racial difference – its 'origins, import, and future implications' (Douglas 2008: 10) – remained as much a topic of debate as the subsumption of particular groups of people under one or the other category. This was especially true for the (non-) distinction of Micronesians from Polynesians. Thus, at the beginning of German colonial rule in the South Pacific in the 1880s, the hierarchical racial division of Oceania was long established.

Today, 'Micronesia' is widely regarded as 'a European invention' (Hardach 2002: 508)[4] whose artificial division had no historical currency among the islanders. During the late nineteenth and early twentieth centuries, however, the alleged racial makeup of the peoples living there remained a matter of anthropological discussion and research. For example, whereas the eminent German physical anthropologist Rudolf Virchow regarded Micronesians as a 'race' apart from Polynesians, his colleague Otto Finsch found himself incapable of distinguishing exact racial boundaries between South Pacific islanders due to the high variability of their physical characteristics. He consequently merely differentiated broadly between Melanesians and 'Polynesians (to which also belong the tricky [*vertrackten*] Micronesians who differ as little from Polynesians as do Swabians from northern Germans)' (1882a: 164).[5] Nonetheless, as I shall show, Finsch postulated a racial hierarchy of Micronesian populations, placing the Marshall Islanders below Nauruans and Gilbertese in terms of both their cultural and physical features.

Finsch, an anthropologist and advocate for the German annexation of South Pacific colonies (in particular New Guinea), travelled around Micronesia and Polynesia between 1879 and 1882. His aim was to collate as much information as possible about the inhabitants of these archipelagos whose assumed imminent extinction provided a justification for the appropriation of their cultural items and physical anthropological data, such as skeletal remains and anthropometric measurements.[6] In the vein of this salvage anthropology[7] – which called for the urgent 'rescue' of their material world rather than their physical existence – Finsch was interested in presumedly 'original' peoples living according to what he regarded as their traditional ways (1882b: 554, 557).[8] He was quickly disenchanted with the inhabitants of the Marshall Islands: incorporating European elements into everyday life, the Micronesian 'original has for the most part been tattered, stained and spoiled by civilization and Christianity' (1892), lamented Finsch.

His visit to the still independent (Hardach 2002: 512) Nauru in July 1880 was his last encounter with such disappointing Micronesians before heading south towards the Melanesian archipelagos. Providing the first German description of Nauruans, Finsch encountered a 'strongly built, beautiful kind of people' (1880a) who appeared to be 'identical' (1882b: 560) to the inhabitants of the Gilbert Islands (Kiribati). Basing his theory on cultural and physical similarities, he suggested they were descendants of early Gilbertese migrants who had no specific features that distinguished them from other Micronesians (Howes 2013: 131).[9] Nevertheless, Finsch declared the Gilbertese 'not only physically but also in many other aspects obviously superior' to Marshall Islanders – most likely because he thought they had not yet 'lost' all of their 'originality' (1882b: 558).[10]

In contrast to their pleasant appearance, he found the Nauruans 'at any rate quite wicked [*ohnehin etwas wüst*]' (1901: 184) with an unstable character and a lack of governance. Finsch detested their intercourse with 'dubious representatives of civilization [which] had less than educational effects on the islanders' (ibid.). These European settler-traders who had lived on the island since the 1840s aggravated the Nauruans' unfriendly nature, providing them with firearms, introducing alcohol and mixing with local women. Eventually, in Finsch's view, the white foreigners' corruptive influences resulted in violent, perpetual tribal warfare; and its destructive effect on their culture made the Nauruans uninteresting for Finsch's further investigation. Instead, he looked forward to parting with 'poor Micronesia' (1882b: 560) and travelled on to the much more exciting New Britain – one of the Melanesian islands inhabited by 'a black population; the home of genuine "man-eaters"' (1880a)[11] living in the 'completely original state' (1882b: 560) more worthy of Finsch's scientific attention.

In 1888, following its contractual annexation two years earlier, the German Imperial Commissioner of the Marshall Islands Protectorate, Franz Sonnenschein, officially incorporated Nauru into Germany's colonial empire. Germany's highest representative in the protectorate was most likely familiar with Finsch's work. He reiterated many of his assessments, describing 'strong and slim-built' Nauruans who stood 'with regard to their physical and mental capacities significantly above' (1889: 21) the Marshall Islanders. Sonnenschein, however, strongly admired what glimpses of Nauruan culture he could catch during his three days' stay. Romanticizing, he perceived the Nauruans as naturally 'childlike and cheerful' (ibid.: 24) but lamented their 'war of each against all' (ibid.). In contrast to Finsch, he attributed their moral decline neither to a naturally wicked disposition nor the influence of unpleasant representatives of European civilization. Instead, Sonnenschein cited a tradition of blood revenge and drought-related hunger periods and put the blame mainly on the Gilbertese for introducing the intoxicating alcoholic drug sour toddy (1889: 22, 24f.).

Given that the Jaluit-Gesellschaft relied on the European traders for the copra trade and the enforcement of its rule, the fact that Sonnenschein exonerated them is not surprising. Quite fittingly, he exerted 'Pax Germanica' (Wedgwood 1936: 360) over the 'big children of Nauru' (Sonnenschein 1889: 25) with the assistance of the two German resident traders by kidnapping their chiefs and releasing them only after their communities handed over their weapons (Hernsheim 1903: 212f.). While it appears plausible that the Nauruans showed little resistance to the implementation of German power because there existed little unity among their communities (Wedgwood 1936: 361), the German colonial narrative maintained that the Nauruans were desperately grateful for this authoritative, external implementation of peace.[12] According to Franz Hernsheim, co-owner of the Jaluit-Gesellschaft, its government forthwith transformed the Nauruans into 'sober, peaceful people who exchange[d] food, clothes and industrial products instead of schnapps and weapons and [were] content with the whole aspect of better living conditions' (1903: 213). Or, as Finsch termed it retrospectively, Germany benevolently 'fulfilled a civilizing task' in Nauru, making it once more 'eligible for its former beautiful name "Pleasant Island"' (1901: 184).

Until the island became a phosphate resource, Nauruans continued to trade copra with the Jaluit-Gesellschaft. Their life seems to have been generally uninterrupted by the stationing of a German government officer on their island supported by two local policemen (Hernsheim 1903: 213; Hardach 2002: 514). Although there were some 'elements on the islands who ha[d] not yet arranged themselves with the new state of things', a succession of resident government officials saw 'little reason to enforce the necessary respect by more drastic intervention' (Jung 1897: 72). It seems that the island officers were mainly engaged in land dispute resolutions (Jung 1897: 71f.; Krämer 1898: 157), and were generally left to gather and report information about the island's nature and inhabitants according to their personal interests. While some of them surveyed the weather or the Nauruan customs, others sporadically undertook censuses by assembling the entire island population on their premises (Steinbach 1896: 547). Whichever information they chose to gather was reported to the German Imperial Commissioner who every now and then travelled from Jaluit in the Marshall Islands to Nauru.

On one such occasion in 1896, the Deputy Officer of the Marshall Island Protectorate, Arno Senfft, romanticized the colonized Nauruans as 'lovely bronze-coloured natives', comparing them with 'elves in a fairy tale landscape' (1896: 103). Describing the men's physique approvingly (ibid.: 104, 105), he distinguished the Nauruans from Melanesians and 'the Micronesian race' by praising their pleasant features and behaviour as 'tending towards the Polynesian race' (ibid.: 103). Replicating the seasoned imagery of the 'noble savage' Polynesian, Senfft portrayed an idyllic, carefree and easy life on Nauru: once the necessary amount of work was cheerfully done, the day was spent playing and chatting away. Through the – self-styled – benevolent nature of Germany's seizure of power, the Nauruans showed 'friendly and modest behaviour towards the Whites' (ibid.: 106). In the same tenor, the ship's doctor and South Sea anthropologist Augustin Krämer concluded his report about his one-day stopover in 1898 by declaring Nauru had been transformed into 'a pretty little island, a German slice of land, an idyll in the Pacific water desert' (1898: 158) inhabited by 'the most peaceful people one can imagine' (1906: 446).

This survey of early German literature about Nauru shows that there prevailed two closely linked themes in early German depictions of Nauruans. The first concerns their predominantly favourable representation as a 'natural' or 'original' people, consistently characterized as physically strong and beautiful. Regarding their disposition, there occurred a shift reflective of the transition in perception from precolonial 'barbarity' to 'colonized noble savagery'. Writing in 1880, Finsch encountered 'wicked' Nauruans corrupted by 'dubious' Europeans. This negative image disappeared with Germany's annexation of Nauru, providing henceforth merely the backdrop for its colonial salvage narrative. According to subsequent German observers of all ranks, the Nauruans were benevolently saved from self-destruction and returned to their traditionally unburdened and peaceful ways of life through Germany's cautious but firm care.

The second theme relates to the racial classification of the Nauruans, which is directly linked to the romanticizing representation of South Sea islanders. Although Nauru was geographically and politically assigned to Micronesia, German observers associated its inhabitants with Polynesians rather than Micronesians. Describing ambiguous characteristics, Germans nonetheless claimed they were culturally and physically superior to supposedly corrupted Marshall Islanders. This alleged racial superiority was explained by suggestions of direct or mixed Polynesian origin. Both of these racial constructions, that is, the shift from wicked to friendly native and the subsequent attaching of the long established ennobling label 'Polynesian', reflect the fact that the 'convoluted schemes' of racial classification referred to above were at the same time stable and adaptable within their respective contexts. As I shall show below, with the beginning of phosphate mining the representation, and with it the treatment of Nauruans, changed again.

Weakening Nauruans

Although Nauruans were frequently depicted as strong and muscular and thus 'would make good workers' (McDaniel and Gowdy 2000: 38), Germans did not comment on their usability for labour before the beginning of small-scale phosphate mining in 1903. Until then, under the Jaluit-Gesellschaft's administration, they continued and were deemed able to produce and trade copra. Whereas Marshall Islanders, for example, were accused of laziness, inaptitude and inefficiency in coconut plantation management, production declines on Nauru were explained by droughts rather than a lack of strength, determination or rationality of its population (Brandeis 1893: 165). Once it was confirmed, however, that their island consisted to a large extent of highly profitable guano deposits and the deal was struck for large-scale phosphate mining, formerly strong Nauruans were deemed physically and mentally weak and thus unfit for work.

In 1903, according to the already mentioned Jaluit-Gesellschaft owner Hernsheim, small amounts of phosphate were shipped in a 'laborious enterprise' (1903: 214) for the first time. While he announced that 'the natives participate willingly in the shipping work because they receive good wages' (ibid.: 215), he left unmentioned the fact that Nauruans must also have extracted the guano deposits. In June 1906, when the PPC's European staff arrived in preparation for large-scale mining operations, a hundred 'Nauru natives ha[d] entered the Company's service' (Gaze 1906). According to Firth, having 'no desire to shovel the phosphate anyway' (1978: 37), they quickly rejected the work as 'arduous and unrewarding' (ibid.: 39). Similar to what Jean Mitchell states about the New Hebrideans' self-determinedness in the ensuing chapter of this volume, the Nauruans were not interested in the capitalist principles of wage labour (McDaniel and Gowdy 2000: 42). They preferred to stay as independent as possible, in control of their land for subsistence, housing, food and the copra production and conditions of trade.

Contemporary Germans explained the Nauruans' disengagement with phosphate mining differently. The German government official on Nauru, Hermann Grössner, took offence that 'the Nauru people only work as long as it suits them; should they certainly take it into their heads to go for a fishing expedition, or if dances and sports occur, the whole party runs away even if they earlier agreed ten times to work' (1906). Notwithstanding such unreliable playfulness, he considered them as potentially fit for work on short-term building projects (ibid.). But the Nauruans were never employed in that capacity; the German engineers responsible for the erection of the mining facilities instead engaged Chinese craftsmen (Ellis 1935: 128).

In contrast to Grössner's culturalist approach to the Nauruans' disinclination to work for the phosphate company, Paul Hambruch offered a more physiological, or rather biological, explanation. One of the main objectives of the Hamburg Südsee-Expedition's leading anthropologist was to scientifically investigate South Sea islander communities with a view to regulate population numbers as a reliable supply for German labour needs in the Pacific tropics (Wiener 1990: 147f.).[13] In 1910, when large-scale phosphate mining was operating on the basis of Chinese and Caroline Islander labour, the PPC welcomed Hambruch on their premises for six weeks to 'work undisturbed with the natives' (Hambruch 1914: v). Familiar with the idealizing publications by Senfft and Krämer (ibid.), he conducted ethnological interrogations and anthropometric investigations. The results of the latter, in Hambruch's view, explained why the Nauruans proved 'useless for physical work' (ibid.: 20). Declaring all previous visitors' impressions of great muscular strength as 'an error' (ibid.: 70), he claimed their 'well proportioned' (ibid.: 69) physique was 'caused by mighty fat deposits. Put to the test through regular work for the phosphate company [they] delivered a sorry result. After a short period of time the worker wore out, and after a few weeks he was unfit for any kind of labour' (ibid.: 70, see also 20 and 56).[14] Hambruch therefore suggested that they should 'train the intelligent potential of the islanders' (1914: 21) for simple work in the administrative and trading sectors.

The Nauruans were never involved in any parts of the phosphate business under German-British company rule (Denoon 1997: 238). When they refused to mine phosphate, their agency in this decision was ignored. Instead, they were declared a physically weak 'race' useless for labour. This weakening of the Nauruans was also undertaken with regard to the re-evaluation of their constitution and it went hand in hand with the colonial-entrepreneurial mismanagement of the Nauruans' health requirements. As I shall show below, the image of the Nauruans was transformed from their depiction as a robust indigenous population into that of a feeble 'race' that was doomed to extinction.

In 1894, the first Marshall Island Protectorate's government physician, Erwin Steinbach, visited Nauru on his return journey to Germany. Steinbach suggested the Nauruans were culturally Micronesians (1896: 549) and physically 'probably a predominantly Polynesian mixed people' (1895a: 161). The physician, however, did not attach much significance to racial categorizations.[15] More concerned about the problems syphilis posed for South Sea islanders, Steinbach presented the Nauruans as a prime example of the Pacific islanders' possible survival from the destructive influences of introduced diseases – under the cautious and protective influence of German colonial rule.

Estimating that fifty per cent of the Marshall Island populations suffered from syphilis (1893: 311), Steinbach found it had a degenerating effect on their physical and mental nature: it rendered the younger generation 'a lot shorter, meagre and weaker' and decreased their 'intellectual stimulative capacity' (ibid.: 312), thus affecting their capability to contribute to the colony's labour force. At the time, such signs of physical and mental demise were interpreted predominantly as evidence for the imminent decline of the respective 'race'. According to the then prevalent 'discourse on the extinction of primitive races' (Brantlinger 2003), the high death rates among indigenous populations caused by introduced diseases and colonial violence were attributed to an alleged lack of adaptability to 'civilization' or mysterious inherent racial weaknesses. Unlike most of his scientist colleagues, however, Steinbach repeatedly refuted such conjecture of doom and saw no reason to fear the demise of South Sea populations (1893: 313; 1896: 547). Instead, he suggested that a combination of acclimatization to the disease (1895b: 470) and the guidance and safeguarding of German health care would 'gradually rebuild them into a healthier and stronger kind of people' (1893: 313).

His claim was based on mere assumptions due to the difficulties of surveying the highly mobile populations in the Micronesian archipelagos (Fitzner 1903: 21).[16] The remote island Nauru, in contrast, offered a quasi laboratory opportunity to observe, if not to stimulate, population growth under competent German rule (Steinbach 1895b: 470f.). Steinbach argued, firstly, after the Germans put an end to the tribal wars, that the Nauruans had returned to their 'completely natural state' (1895a: 161), allowing for a 'natural rate of die-off' (ibid.: 162). Secondly, the Nauruans had escaped the devastating consequences of syphilis because the Germans '[kept] the natives in their own interest in a kind of artificial isolation' (1896: 549). Steinbach's evidence consisted of the two censuses undertaken in 1889 and 1893, according to which the population had increased even despite Nauru's recurring drought-related periods of malnutrition (which at the time of Steinbach's three-day visit left the Nauruans 'physically a bit deteriorated' [ibid.: 547]). Thus the combination of maintaining a so-called 'natural' way of life, acclimatization and racial isolation under German administration, according to Steinbach, provided the conditions for healthy, growing populations in the rest of Micronesia.

Steinbach was the first of a series of government physicians stationed in Jaluit as part of the agreement between the German government and the Jaluit-Gesellschaft. But the company was generally unwilling to arrange and pay for the health care of islanders in the Pacific colonies (Eckart 1988: 97).[17] The mismanagement of diseases breaking out in the early 1900s on Nauru reflects the gruesome reality that German-British 'government by company' had less concern for the welfare of Pacific islanders than it claimed. The phosphate mining companies certainly had no time for Steinbachian acclimatization processes (Firth 1978: 37; Eckart 1988: 93).

Prior to the PPC's establishment in 1906, the German government physician for the Marshall Islands Protectorate (for example, Steinbach) irregularly visited Nauru for the medical inspection of its inhabitants. In general, however, the Catholic and Protestant missions provided and paid for the medical treatment of the Nauruans (Verhandlungen des Reichstages 1907a: 361f.). This practice was replaced by the PPC's employment of a German resident doctor who was predominantly in charge of its employees and the medical inspection of incoming labourers. He also took on the responsibilities of government physician, caring for non-company related island inhabitants (Gaze 1907f; Steudel 1909: 922; Eckart 1988: 92).

Treating the Nauruans in this double role, the initial resident doctor Hermann Walbaum came into conflict with the phosphate company's German manager on Nauru, Emil Larsen. Larsen not only criticized Walbaum's medical supply orders as unnecessary but also questioned his treatment of Nauruans at the company's expense (Gaze 1907a, 1907b, 1907e). Although the PPC agreed to bear the costs (Gaze 1907g), the Nauruans in reality had to pay for their medical treatment with a hefty annual copra tax (Walbaum 1908: 256; Eckart 1988: 92). As I shall elaborate further below, the conflicts between doctor and manager also concerned the treatment of the Chinese labourers and quickly led to the replacement of both. Walbaum eventually quit, leaving his post to Adolf Müller (Gaze 1907h) who stayed on as doctor until the island was seized by Australian troops in 1914 (Eckart 1988: 92).

According to Eckart, the newly appointed Müller 'did not dare to criticize' (ibid.: 96) the phosphate company after the sacking of his predecessor and therefore stuck to recording statistics. From his study on German health care in the Pacific Eckart has also concluded that German government physicians 'tried to do what they could in providing the native population with medical care and protecting the company's workers against ruthless exploitation' (ibid.). On Nauru, this might have been true for Walbaum, who was apparently well liked by the Nauruans (Gaze 1907c) and whose criticism of the PPC's handling of devastating epidemics among its Chinese workers probably led to his resignation (Eckart 1988: 95). The same can hardly be said of Müller, whose medical publications contained in fact much more than numbers, especially regarding assumptions that linked health to 'race'; namely, a connotative shift in the meaning of 'ignorance', from describing a happy, carefree and healthy people to blaming the Nauruans for their illnesses due to carelessness and a lack of hygiene and, eventually, for their inevitable decline.

Throughout roughly the first decade of German administration, an increase in population numbers appeared to confirm Steinbach's optimistic outlook (Fitzner 1903: 24–25; Hermann 1908: 561–562); thus there seemed to be little need for concern for the Nauruans' health. In 1903, Hernsheim enthusiastically asserted that Nauru was 'completely free of fever' (211) and therefore cost the Reich not a penny, although more or less severe influenza infections regularly troubled the Nauruans (Walbaum 1908: 259). But only two years later, and for the first time since German administration, the irregularly conducted censuses registered more deaths than births due to the outbreak of dysentery (Verhandlungen des Reichstages 1907b: 88; Hermann 1908: 562). By July 1907, another dysentery epidemic again killed many islanders, followed by a whooping cough infection that affected nearly every Nauruan (Müller 1909: 482).

The exact death toll among Nauruans caused by these epidemics of 1905 and 1907 remains unclear, due to inconsistent and racially selective and generalizing recording. In the years 1905 to 1908 the population statistics for the Marshall Islands presented to the German Reichstag differentiated 'the white population' from the 'non-native coloured population and mixed-bloods according to race and sex' but did not include information about the protectorate's indigenous populations. Only for the period 1908 to 1909 did the Nauruans appear under the category 'Native Population' (Verhandlungen des Reichstages 1911: 944f.). But, while the total population numbers for Nauruans were recorded for each of those years in the textual part of the reports, the actual causes of death were registered exclusively for Europeans (idem 1907c: 357–360; 1908: 4133–137; 1909c: 7077–7081; 1911: 944–945).

Regarding 1905, the Reichstag report stated that the health situation on Nauru was 'not good' as dysentery and influenza 'caused many deaths among the natives' (idem 1907d: 88). The Protestant mission on Nauru also reported that 'a lot of sickness governed the island', stating there were 'many deaths' (idem 1907a: 361) in 1905, three of which occurred in the mission station. In total, there were sixty-eight deaths – more than double the number of previous years (Verhandlungen des Reichstages 1907d: 88; Hermann 1908: 562). While the higher death rate was implicitly linked to the epidemics (Verhandlungen des Reichstages 1908: 4125; Hermann 1908: 562), it remains unclear exactly how many died of dysentery and influenza. Therefore, in 1905, it can be assumed that around thirty Nauruans died from dysentery, influenza or a combination of both.

The records of the 1907 epidemics are even more ambiguous. In the wake of influenza and dysentery Walbaum recorded 'several cases of deaths' (1908: 259) when he left the island in June 1907. By the time Müller reported on Nauru's health situation, he had treated fifty-one Nauruans with 'dysentery [which] claimed many victims' (Verhandlungen des Reichstages 1909a: 4125; Müller 1909: 495) but, unlike for Europeans, Chinese and Caroline Islanders, he gave no indication of how many died (see Figure 2.2). The Catholic mission reported that sixty-four of its converts died from 'a variety of epidemically occurring illnesses' (idem 1909b: 7169). Firth has stated that seventy to eighty Nauruans had died by the end of June 1907 (1978: 41), but it is probable that the number was much higher. According to local knowledge 'dysentery carried off about 150 people' (Wedgwood 1936: 369) – that would mean about a tenth of the entire population.

While Walbaum reported that the dysentery epidemic originated in the phosphate company's labourer quarters where the recently arrived workers were also affected by it (1908: 259), Müller blamed the Nauruans for their suffering. He opened his first annual medical report by stating that 'the spread of any disease among the natives is facilitated by their ignorance and uncleanliness' (1909: 482). This general accusation, which contradicted some of his predecessors' informative descriptions of, for example, Nauruan indigenous medical expertise, general neatness and cleanliness (see e.g. Finsch 1880a; Krämer 1898: 154), provided the PPC's

Jahresnachweisung über die kranken Einwohner der Station Nauru. 495

Lfde. Nr.	Krankheit	Bestand waren	Zugang	Im ganzen behandelt	Abgang: geheilt	Abgang: gestorben	Abgang: anderweitig	Abgang: Summe	Bestand bleiben
	XI. Krankheiten der Bewegungsorgane.								
30	Rheumatismus musculorum	.	11	11	11	.	.	11	.
	Summe XI	.	11	11	11	.	.	11	.
	XII. Mechanische Verletzungen.								
31	Weichteilwunden	.	9	9	9	.	.	9	.
32	Kontusion	.	2	2	2	.	.	2	.
33	Distorsion	.	1	1	1	.	.	1	.
34	Komplizierte Fraktur	.	1	1	.	.	.	.	1
35	Quetschung mit Zehengangrän	.	1	1	1	.	.	1	.
.	Summe XII	.	14	14	13	.	.	13	1

D. Nauru-Eingeborene.

Es wurden behandelt an:

	I. Infektionskrankheiten usw.	
1	Influenza	3
2	Typhus abdominalis	1
3	Dysenterie	51
4	Keuchhusten	55
5	Tuberkulosis	4
6	Akuter Gelenkrheumatismus	5
7	Frombösie	72
8	Atrophie	1
9	Pseudoleukämie	1
	Summe I	193
	II. Krankheiten des Nervensystems.	
10	Kopfschmerzen	2
11	Migräne	1
12	Neuralgie	3
13	Schlaflosigkeit	1
14	Eclampsia infantum	1
	Summe II	8
	III. Krankheiten der Atmungsorgane.	
15	Bronchitis	65
16	Pneumonia crouposa	5
17	Bronchopneumonie	4
18	Pleuritis	6
19	Asthma bronchiale	12
20	Emphysema pulmon.	1
	Summe III	93

	IV. Krankheiten der Kreislaufs- usw. Organe.	
21	Vitium cordis	1
22	Stenokardie	1
23	Lymphadenitis	5
24	Drüsentumoren	1
25	Drüsentuberkulose	1
	Summe IV	9
	V. Krankheiten der Ernährungsorgane.	
26	Überflüssige Zähne	2
27	Caries dentium	12
28	Stomatitis	4
29	Aphthen	1
30	Parulis	1
31	Parotitis	1
32	Eiterige Entzündung der Sublingualis	1
33	Dyspepsie	19
34	Gastralgie	8
35	Gastroenteritis acuta	13
36	Enteritis acuta	18
37	Brechdurchfall	44
38	Diarrhöe	5
39	Obstipatio	3
	Summe V	132

Figure 2.2. Adolf Müller's 'Annual report on the sick inhabitants of the station Nauru' (1909: 495) showing the number and causes of deaths for phosphate miners (here Caroline Islanders) and Nauruans treated for illnesses between 2 July 1907 and 31 March 1908

Australian representative, Harold Gaze, with a convenient argument against his concerned employers in London. As will become clear below, the PPC's London board signalled great concern about the ill health of both Nauruans and Chinese workers, sending out repeated requests and propositions to ameliorate the situation. Reiterating Müller's blame-the-patient approach, Gaze was quick to assert that '[t]he mortality among the Natives has no special relation to the ill-health of the Coolies, but that the two arose from quite different causes'. To him there was '[n]o doubt the large number of death among the natives ... were caused by the epidemic of Dysentery and the lack of sanitary precaution on the part of the natives' (Gaze 1907l).

Müller's disdainful view of the Nauruans as primitive and ignorant became again apparent in an article about a short but severe epidemic of a 'peculiar and hitherto unknown disease' (1910b: 535) which swept through the non-Chinese population in January 1910. 'Neuromyelitis infectiosa' (ibid.: 543; probably infantile paralysis [Wedgwood 1936: 369]) caused fever, pain and paralysis of the limbs; in severe cases sufferers died from its effect on the respiratory and digestive systems. Within a fortnight, 470 Nauruans (that is, more than a third of the population) and 220 Caroline Islanders became ill; thirty-seven Nauruans and one Caroline Islander died a 'quite agonising' (1910b: 541) death. While Müller (apparently unwittingly) documented in his article that the Nauruans were in fact very capable of describing their symptoms (ibid.: 537–538), he credited merely 'the more intelligent' (ibid.: 538) of them with accurately identifying their decreasing nerve sensitivity and assigned 'value' (ibid.) only to the information given by one of the four mildly affected Europeans. Acknowledging that the infectious disease's variable symptoms and course made its cause indeterminable, Müller dismissed the possibility of its introduction. Again, he blamed the Nauruans' 'unhygienic habits' (ibid.: 542), insinuating they were ignorant towards his medical expertise.[18]

At the time, the third year of a severe drought gripped Nauru (Müller 1911: 546; Hambruch 1914: 51, 57). It destroyed most coconut plantations, causing famine and exhaustion, which would clearly have aggravated the effects of any kind of illness. In his report to the German government, Müller acknowledged that 'the natives' nutrition naturally suffered' to such a degree that they 'even hired themselves out as labourers, which surely could only happen in dire strait' (1911: 546).[19] Their will to survive probably explains why those who still showed serious signs of paralysis three months after the end of the epidemic nevertheless strove to return quickly to their daily business, even if they needed support with their walking. In his medical publication (aimed at his colleagues and thus a means to gain academic standing) Müller did not even mention the drought and represented this behaviour as a sign of the Nauruans' irrational, 'quite indifferent' (1910b: 541) attitude towards life in general. Further, he alleged that some died 'without an objective cause' from mysterious machinations, through which 'with the expiry of the will to live the vitality too seems to come to an end' (ibid).

Müller's implicit reference to supposedly innate 'race' characteristics, which rendered 'the colonized' incapable of surviving contact with 'civilization' and foreign 'races', reiterated the virulent contemporaneous European phantasy of doomed 'primitive races' in the South Pacific. This trope had already motivated Finsch to visit Micronesia; and Steinbach subsequently rejected the theory on the basis of the Nauruan example. This extinction discourse served to justify European colonial appropriation and exploitation of lands and humans by relegating the responsibility for the mostly human-made and avoidable population declines to the victims.[20] Such ideas about racial feebleness also influenced the treatment of the Chinese indentured labourers in Nauru's phosphate mines.

Exploiting the Chinese

Even if the Nauruans had been more amenable to phosphate mining, they were far too few to satisfy the large labour demand of early twentieth-century phosphate extraction. An alternative labour source was essential and the indenture of labourers was arranged through Germany's colonial enclave in China, Kiautschau (Ellis 1935: 128; Firth 1978: 37, 39). But, to the PPC's dismay, construction work did not begin smoothly when the first Chinese tradesmen finally arrived on Nauru, due to their unmistakable readiness to go on strike because of injustices committed against them (Firth 1978: 39–40). The first 438 labourers arriving in March 1907 proved difficult to manage.[21] Having been deceived about the level of their wages, they too went on strike (Fitzgerald 2007: 174). In response to attempts to starve the strikers, they looked for food in the Nauruans' settlements and caused havoc on their island.

Figure 2.3. 'Nauru. Chinese coolies at work in the phosphate fields' (Verhandlungen des Reichstages 1911: 963)

Eventually, nearly half of the Chinese, insisting on their promised wages, were shipped back (Firth 1978: 40) labelled 'unusable' (Müller 1909: 483).

Far from acknowledging the Chinese's solidarity as a self-determined response to unacceptable employment conditions, the manager Larsen declared them as 'not a very satisfactory lot' (Gaze 1907d). Coming from the colder parts of north China, they were unsuitable for work in the tropics (idem 1907a) and Larsen only hoped that the next shipment would be 'superior to the first' (idem 1907j).

They were indeed incapacitated, because dysentery and typhoid fever broke out among the Chinese workers. Many of them also suffered from beriberi, a vitamin B1 deficiency caused by a diet based on de-husked or polished rice, that would kill horrific numbers of Chinese labourers on Nauru. As I have mentioned above, in Müller's initial report to the Colonial Office (covering the period from his arrival on Nauru on 22 July 1907 until 31 March 1908), he differentiated between Europeans, Chinese, Caroline Islanders and Nauruans. In the following years, however, Müller subsumed all non-Europeans under the category 'Coloureds' [*Farbige*]. This generalization compounds the problem of ascertaining exact numbers for treated and deceased Nauruans, Caroline Islanders and Chinese – except for cases of beriberi, which only occurred among the Chinese. Because of this, I shall focus here on these outbreaks of beriberi.

While beriberi was common among Asian labourers in European colonies, its origin was unknown at the time. Medical researchers had found evidence for a connection to rice consumption, but explanations of the disease ranged from general malnutrition to a mysterious infection (Nocht 1920; Thurnham 2005: 269–270). That lack of knowledge made treatment difficult for Walbaum who saw the cause in preventable poor water and food quality and criticized the PPC harshly for its negligence towards its Chinese labourers (Eckart 1988: 95). His attempts to improve their rations were thwarted by Larsen (Firth 1978: 41), whose sabotage of the physician's treatment efforts appears to have stemmed not only from his disregard for the Chinese but also from conflicts which raged between him and the other Germans working for the PPC. These difficulties were eventually resolved by both Larsen's and Walbaum's replacement in mid-1907 (Ellis 1935: 138).[22]

Walbaum noted that at the end of March 1907 several cases of beriberi had occurred among the approximately 230 Chinese labourers[23] and at the beginning of July, Gaze had to report repeatedly to his London board that the Chinese were 'dying at a high rate' (1907i). By the middle of the month, thirty-nine had perished (idem 1907j). While it remains unclear exactly how many suffered and died from this first beriberi epidemic, Müller claimed it was over when he arrived on 22 July 1907, leaving him with ten patients (1909: 483, 489).

In early November 1907, 'around 390' (ibid.: 483) new Chinese workers arrived in Nauru. Initially regarded as 'a very capable lot' (Gaze 1907m), it soon turned out that 123 of the newly arrived Chinese also suffered from beriberi. Only thirty regained their health, sixty-six died and thirty-six were still ill at the end of March 1908. An additional 148 workers contracted dysentery of whom seventeen did not survive; and typhoid fever claimed another fourteen Chinese (Müller 1909: 489).

According to Müller, this second beriberi episode began to ebb from March 1908, but a third beriberi affected group of Chinese tradesmen arrived in July (1910a: 453). Affecting 181 workers, the disease peaked in October and again left sixty-six dead. Thirty-one were still sick at the end of March 1909 (ibid.: 459). The fourth wave hit in the following year; this time, of the 112 who suffered from beriberi only thirteen Chinese died (idem 1911: 552) – according to Müller because they were mostly 'relapses' (ibid.: 547).

The PPC's London board worried about the 'very distressing' (Gaze 1908a) news concerning the Chinese's 'alarming death rate' (idem 1908b). Although Gaze's employers frequently suggested a correlation between the severity of the Chinese's decline and their living conditions, he proved unwilling to accept such a causal relation. For example, in September 1907, London suggested that the high mortality rate might be linked to the housing near the beach. Dismissing the possibility of dampness in their accommodation, Gaze pronounced that it did not have 'an important influence on the number of deaths' (1907k) and continued to dismiss proposals to change the accommodation, provision and sanitary conditions (which had even been criticized by Hernsheim), insisting everything was being done on Nauru to deal with the deadly disease (1908e).

Gaze was corroborated by Müller, who had cleared the majority of the new arrivals for work in the mines. Asserting that neither the 'food theory' (Müller 1909: 484)[24] nor the accommodation could explain the presence of beriberi, he now claimed that many of the sick had already arrived in a 'weakly' (ibid.) condition. Gaze argued in the same vein, purporting that certain groups of the Chinese were simply unsuited to the Nauru climate. This time however, it was those from the south rather than the north, whom Larsen had previously deemed unable to withstand the island's seasonally damp weather (Gaze 1908b). Unless a 'better class of men from the north of China' (idem 1908c) were acquired, Gaze wrote, the eventual replacement of Chinese by Caroline Islanders must be contemplated (ibid., idem 1908d). Unhappy about the mistreatment of its citizens by the German Empire and its associates, it was in fact the Chinese government rather than the PPC that put a stop to the indenture of Chinese labourers in the same year (Fitzgerald 2007: 174–176).

As Firth has rightly stated, even though the workers introduced beriberi to Nauru, the PPC 'must share the blame for the appalling death toll among the Chinese' (1978: 41). Obviously the cause of the deadly condition was not fully known but it is also clear that the way in which it was managed on the island was significantly influenced by perceptions of 'race', which materialized, firstly, in the stubborn resistance on the ground in Nauru and in Melbourne in the face of the London headquarters' enquiries and, secondly, in the rejection of responsibility for what was going on in the labourer quarters. Those responsible did not change the living conditions, which facilitated the spread of dysentery, nor did they manage to provide the workers with better food. As far as one can tell, the company's board members were more inclined to find a solution to the health problems on Nauru than their staff, be it only to ensure the smooth, that is, profitable running of its mining operations.

Conclusion

This excursion into the complexities of 'race', health and labour on Nauru during German administration can only spotlight some of the many dynamics of colonial entrepreneurial governance on the island. As Mitchell will show in the following chapter of this volume, in the context of colonial health management, changes in the perception of indigenous populations occurred according to the political and social context. In the German colonial mind-set, the Nauruans were returned to their alleged natural happy existence through German administration, but this image of healthy noble savages changed with the beginning of phosphate mining. The Nauruans' rejection to work in the mines could then only mean that they were culturally and racially unfit, rather than self-determinedly unwilling, to excavate phosphate.

When large-scale mining was set to begin in 1907, diseases such as dysentery and beriberi in the Chinese labour quarters endangered the smooth running of the enterprise. The spreading of dysentery among the Nauruans was just as troubling, at least to the faraway head office directors in London. The local colonial and entrepreneurial representatives, however, rejected any responsibility for the deteriorating health and the high death rates among both the indigenous and indentured populations of Nauru. They preferred to associate them with their allegedly ill-suited 'race' nature.

In the case of the Nauruans, it was claimed that their ignorance, unhygienic lifestyle and their essential incapacity to withstand the contact with foreign 'races', was to blame. With regard to the Chinese, the similarly obvious causes for their suffering and dying were also relegated to the realm of racial nature. Rather than changing their unbearable, eventually deadly, living conditions – imposed upon them through the rendering of humans into cheap labour resources – it was suggested the quality of their 'race' was inferior or not suited for the Nauruan climate.

This denial of responsibility for the well-being and survival of the indigenous and indentured populations on Nauru was also caused by divisions on an individual and structural level. Altogether, the personal feuds between Walbaum and Larsen, the different attitudes towards the Nauruans and Chinese shown by Walbaum and Müller, and the apparent divide between the PPC's London headquarters and their representatives on the ground all point to the impact individuals and structures have within the unequal power relations that exist in the colonial sphere.

Acknowledgements

I am grateful to the editors of this volume, Alexandra Widmer and Veronika Lipphardt, for their encouraging comments on early drafts of this chapter. Many thanks also go to the peer readers of the final draft, whose insightful suggestions I found very valuable.

Antje Kühnast has submitted her doctoral thesis (History and Philosophy of Science, School of Humanities & Languages at the University of New South Wales, Sydney) on the appropriation and scientific utilization of Australian Aboriginal skeletal remains by German physical anthropologists during the nineteenth century. Her research interests include the histories of racial thought and theories of human evolution, racism analysis and the history of physical anthropology, especially in the German-language sphere. She has published two book chapters in those fields.

Notes

1. See also Ellis (1935: 45f.).
2. See also Eckart (1988: 96) and McDaniel and Gowdy (2000: 35).
3. The original papers of the Pacific Phosphate Company are located in the National Archives of Australia. I consulted the microfilm copies available at the Mitchell Library in Sydney through the Pacific Manuscript Bureau (PMB) from http://asiapacific.anu.edu.au/pambu/ (AU PMB MS1207 Reels 4 and 5).
4. All translation from German to English are my own.
5. For a comprehensive analysis of Finsch's views on race in Oceania, see Howes (2013).
6. The visit to Nauru presented a potential opportunity for Finsch to acquire skeletal material for the German leading physical anthropologist Rudolf Virchow. To Finsch's regret he was unable to obtain a single skull 'although some thirty natives had been slaughtered during the last three months'. He thus had to content himself with obtaining only a variety of ethnological items (Finsch 1880b). I would like to sincerely thank Hilary Howes for forwarding this reference. In his 1901 publication Finsch revised the number of 'slaughtered natives', stating that the Nauruans 'boasted in probably exaggerating fashion that during the last three to four months 35 "warriors" had fallen' (Finsch 1901: 184). He thus questioned the accuracy of the Nauruans' statements and played down both the intensity of their warfare and its possible impact on future population developments.
7. On German salvage anthropology in the South Pacific, see Buschmann (2007).
8. See also Howes (2013: 13, 255–259).
9. In 1880, shortly after his visit to Nauru, Finsch based this assumption on similarities of skin shade. But he neutralized language as a defining factor for racial classification, noting that their languages differed so much that Gilbertese and Nauruans were unable to understand each other (see Finsch 1880a, 1882b). By 1901, he reassessed this view on linguistic significance, stating that the Nauruans were 'physical anthropologically and ethnologically very closely linked to the Gilbertese' and spoke a language only slightly different from the Gilbertese's (1901: 184).
10. See also Howes (2013: 256).
11. Two years later Finsch repeated this sentiment but omitted the quotation marks around the term 'man eaters' (1882b: 560). On Melanesia as the assumedly most important region for German late nineteenth-century anthropology, see Buschmann (2007: 305–307).
12. See, together with Finsch's representations of the Nauruan cited above, the contemporaneous accounts by Steinbach (1896: 548–549), Senfft (1896: 103), Jung (1897: 72), Hambruch (1914: 13–17). But this narrative is frequently reiterated even today, see e.g. the website of the Deutsch-Nauruische Gesellschaft, retrieved 20 September 2015 from http://www.nauru.de/fakten_ueberblick.php, Längin (2004: 243). It has also been

claimed that Germany saved the Nauruans from 'self-destruction' (Folliet 2011: 24; McDaniel and Gowdy 2000: 35).

13. For more detail on the expedition, see Fischer (1981).
14. The island doctor Adolf Müller too alluded to the 'spongy corpulence' of the Nauruans 'in contrast to the much more agile and sinewy Truk people' who worked in the phosphate mines (1910b: 540).
15. Steinbach rarely used the German term for 'race' [*Rasse*] to define the peoples of the Pacific region; instead he referred to Oceanic peoples as tribes [*Volkstämme*] or kinds of people [*Menschenschläge*].
16. This was also stated regularly in the health reports to the German government about the Marshall Island Protectorate.
17. On the Jaluit-Gesellschaft's general reluctance to pay for the medical treatment of Marshall Islands communities, see Eckart (1988: 93–94).
18. Müller repeated this accusation in the health report of the following year (1910a: 453).
19. Paul Hambruch, the anthropologist of the 1910 Hamburg Südsee-Expedition, however, reported that the phosphate company saved the Nauruans from starvation by supplying them with food rations; see Hambruch (1914: 51).
20. Eckart refers to the German variant of extinction discourse as 'Aussterbe-Mechanismus (mechanism of extinction)', (1988: 85).
21. According to Firth they numbered 457 (ibid.: 40).
22. According to Eckart, Walbaum was dismissed 'through the influence of the Pacific Phosphate Company' following a letter of complaint about the living conditions of the Chinese labourers (1988: 95).
23. According to Firth, of the 457 Chinese, 221 were sent back following their strike, thus 236 remained (1978: 40). Gaze related on 3 July 1907 that 220 Chinese workers were on Nauru (1907i); Müller stated that there were 'around 300', probably including the Chinese tradesmen that had arrived earlier (1909: 483).
24. Müller obviously believed beriberi was an infectious disease, as he assumed that one tradesman arriving in July 1908 'became sick from a relapse [brought in] from China and from this the third epidemic on the island followed' (1910a: 453). In 1911 he mentioned that a Nauruan had contracted beriberi in the prison where a number of Chinese workers were also incarcerated (1911: 547).

Bibliography

Brandeis, E. 1893. 'Bericht, betreffend das Schutzgebiet der Marschallinseln, für das Jahr 1891', in *Stenographische Berichte über die Verhandlungen des Reichstages. 8. Legislaturperiode. 2. Session 1892/93. Erster Anlageband.* Berlin: Norddeutsche Buchdruckerei, pp. 164–166. Retrieved 20 September 2015 from http://www.reichstagsprotokolle.de/Blatt3_k8_bsb00018683_00199.html

Brantlinger, P. 2003. *Dark Vanishings. Discourse on the Extinction of Primitive Races, 1800-1930.* Ithaca: Cornell University Press.

Buschmann, R.F. 2007. 'Oceanic Carvings and Germanic Cravings', *Journal of Pacific History* 42(3): 299–315.

Denoon, D. 1997. 'New Economic Orders: Land, Labour and Dependency', in D. Denoon, M. Meleisea, S. Firth and J. Linnekin (eds), *The Cambridge History of the Pacific Islanders.* Cambridge: Cambridge University Press, pp. 218–252.

Douglas, B. 2008. 'Foreign Bodies in Oceania', in C. Ballard and B. Douglas (eds), *Foreign Bodies. Oceania and the Science of Race 1750-1940.* Canberra: Australian National University Press, pp. 3–30.

Eckart, W.U. 1988. 'Medicine and German Colonial Expansion in the Pacific: The Caroline, Mariana and Marshall Islands', in R. MacLeod and L. Milton (eds), *Disease, Medicine, and Empire.* London: Routledge, pp. 80–102.

Ellis, A.F. 1935. *Ocean Island and Nauru: Their Story.* Sydney: Angus and Robertson.

Finsch, O. 1880a. 'Nawodo (Pleasant Island)', *Hamburger Nachrichten,* 1 December, [no pages].

—— 1880b. Letter to Rudolf Virchow, 26 August, Nachlass Virchow, Nr. 607, Berlin-Brandenburgische Akademie der Wissenschaften.

—— 1882a. 'Die Rassenfrage in Oceanien', *Verhandlungen der Berliner Gesellschaft für Anthropologie, Ethnologie und Urgeschichte* 14: 163–166.

—— 1882b. 'Ueber seine in den Jahren 1879 bis 1882 unternommenen Reisen in der Südsee', *Verhandlungen der Gesellschaft für Erdkunde zu Berlin* 10: 553–564.

—— 1892. Letter to Unknown, 25 March, cited by Howes (2013: 256).

—— 1901. 'Der Fregattvogel und sein Fang auf Nauru', *Velhagen & Kalings Monatshefte* 15(8): 178–184.

Firth, S. 1973. 'German Firms in the Western Pacific Islands, 1857-1914', *Journal of Pacific History* 8: 10–28.

—— 1978. 'German Labour Policy in Nauru and Angaur, 1906-1914', *Journal of Pacific History* 13(1): 36–52.

—— 1982. *New Guinea Under the Germans.* Melbourne: Melbourne University Press.

Fischer, H. 1981. *Die Hamburger Südsee-Expedition.* Frankfurt a. M.: Syndikat.

Fitzgerald, J. 2007. *Big White Lie: Chinese Australians in White Australia.* Sydney: University of NSW Press.

Fitzner, R. 1903. 'Die Bevölkerung der deutschen Südseekolonien', *Globus* 84(2): 21–25.

Folliet, L. 2011. *Nauru. Die verwüstete Insel.* Berlin: Verlag Klaus Wagenbach.

Gaze, H. 1906. Letter to Secretary London Office, 18 July 1906, AU PMB MS1207 Pacific Phosphate Company Co. Ltd. Sydney and Melbourne Offices: London Correspondence, 1902-1923. Reel 4 MP1174/1/15 PPC Sydney. Letters and Enclosures from Sydney Office to PPC London, 'General re Nauru' no. 140.

—— 1907a. Letter to Secretary London Office, 9 January 1907, AU PMB MS1207 Reel 4 MP1174/1/15 PPC Sydney. Letters and enclosures from Sydney Office to PPC London, 'General Nauru' no. 195.

—— 1907b. Letter to Secretary London Office, 30 January 1907, AU PMB MS1207 Reel 4 MP1174/1/15 PPC Sydney. Letters and enclosures from Sydney Office to PPC London, 'General Nauru' no. 201.

—— 1907c. Letter to Secretary London Office, 6 February 1907, AU PMB MS1207 Reel 4 MP1174/1/15 PPC Sydney. Letters and enclosures from Sydney Office to PPC London, 'General re Nauru' no. 202.

—— 1907d. Letter to Secretary London Office, 6 March 1907, AU PMB MS1207 Reel 4 MP1174/1/15 PPC Sydney. Letters and enclosures from Sydney Office to PPC London, 'General re Nauru' no. 217.

—— 1907e. Letter to Secretary London Office, 3 April 1907, AU PMB MS1207 Reel 4 MP1174/1/15 PPC Sydney. Letters and enclosures from Sydney Office to PPC London, 'General Nauru' no. 233.

——— 1907f. Letter to Secretary London Office, 10 April 1907, AU PMB MS1207 Reel 4 MP1174/1/15 PPC Sydney. Letters and enclosures from Sydney Office to London, 'Finance and Accounts' no. 688.

——— 1907g. Letter to Secretary London Office, 17 April 1907, AU PMB MS1207 Reel 4 MP1174/1/15 PPC Sydney. Letters and enclosures from Sydney Office to PPC London, 'General Nauru' no. 239.

——— 1907h. Letter to Secretary London Office, 22 May 1907, AU PMB MS1207 Reel 4 MP1174/1/15 PPC Sydney. Letters and enclosures from Sydney Office to PPC London, 'General re Nauru' no. 261.

——— 1907i. Letter to Secretary London Office, 3 July 1907, AU PMB MS1207 Reel 5 MP1174/1/16 PPC Melbourne. Letters from Melbourne to PPC London, 'General re Nauru' no. 285.

——— 1907j. Letter to Secretary London Office, 17 July 1907, AU PMB MS1207 Reel 5 MP1174/1/16 PPC Melbourne. Letters from Melbourne to PPC London, 'General re Nauru' no. 293.

——— 1907k. Letter to Secretary London Office, 18 September 1907, AU PMB MS1207 Reel 5 MP1174/1/16 PPC Melbourne. Letters from Melbourne to PPC London, 'General re Nauru' no. 331.

——— 1907l. Letter to Secretary London Office, 2 October 1907, AU PMB MS1207 Reel 5 MP 1174/1/16 PPC Melbourne. Letters from Melbourne to PPC London, 'General re Nauru' no. 341.

——— 1907m. Letter to Secretary London Office, 27 November 1907, AU PMB MS1207 Reel 5 MP1174/1/16 PPC Melbourne. Letters from Melbourne to PPC London, 'General re Nauru' no. 372.

——— 1908a. Letter to Secretary London Office, 11 March 1908, AU PMB MS1207 Reel 5 MP1174/1/16 PPC Melbourne. Letters from Melbourne to PPC London, 'General re Nauru' no. 431.

——— 1908b. Letter to Secretary London Office, 22 April 1908, AU PMB MS1207 Reel 5 MP1174/1/16 PPC Melbourne. Letters from Melbourne to PPC London, 'General re Nauru' no. 457.

——— 1908c. Letter to Secretary London Office, 13 May 1908, AU PMB MS1207 Reel 5 MP1174/1/16 PPC Melbourne. Letters from Melbourne to PPC London, 'General re Nauru' no. 469.

——— 1908d. Letter to Secretary London Office, 5 May 1908, AU PMB MS1207 Reel 5 MP1174/1/16 PPC Melbourne. Letters from Melbourne to PPC London, 'General re Nauru' no. 465.

——— 1908e. Letter to Secretary London Office, 19 May 1908, AU PMB MS1207 Reel 5 MP1174/1/16 PPC Melbourne. Letters from Melbourne to PPC London, 'General re Nauru' no. 474.

Great Britain. Parliament. House of Commons. 1886. *Declaration Between the Governments of Great Britain and the German Empire Relating to the Demarcation of the British and German Spheres of Influence in the Western Pacific, Signed at Berlin, April 6*. Retrieved 20 September 2015 from http://marshall.csu.edu.au/Marshalls/html/history/UKTreaty1.html

Grössner, H. 1906. Letter to Voss, 24 February 1906, AU PMB MS1207 Reel 10 MP1174/1/13 PPC Melbourne. Enclosures from London no. 133.

Hambruch, P. 1914. *Nauru. 1. Halbband. Ergebnisse der Südsee-Expedition 1908-1910.* Hamburg: L. Friedrichsen & Co.

Hardach, G. 2002. 'Deutsche Herrschaft in Mikronesien', in H.J. Hiery (ed.), *Die Deutsche Südsee*, 2nd edn. Paderborn: Schöningh, pp. 508–534.

Hermann, R. 1908. 'Zur Statistik der Eingeborenen der deutschen Südseegebiete', *Zeitschrift für Kolonialpolitik, Kolonialrecht und Kolonialwirtschaft* 11: 556–568.

Hernsheim, F. 1903. 'Nauru, eine gehobene Koralleninsel der Marshall-Gruppe', *Mitteilungen der Geographischen Gesellschaft in Hamburg* 19: 211–215.

Howes, H. 2013. *The Race Question in Oceania. A.B. Mayer and Otto Finsch Between Metropolitan Theory and Field Experience, 1865-1914*. Frankfurt a.M.: Peter Lang.

Jung, F. 1897. 'Aufzeichnungen über die Rechtsanschauungen der Eingeborenen von Nauru', *Mittheilungen von Forschungsreisenden und Gelehrten aus den Deutschen Schutzgebieten* 10: 64–72.

Krämer, A. 1898. 'Nauru. Ein Besuch der Insel', *Globus* 74(10): 153–158.

——— 1906. *Hawaii, Ostmikronesien und Samoa: Meine Zweite Südseereise (1897-1899) zum Studium der Atolle und ihrer Bewohner*. Stuttgart: Strecker & Schröder.

Längin, B.G. 2004. *Die deutschen Kolonien. Schauplätze und Schicksale 1884-1918*. Hamburg: Verlag E. S. Mittler & Sohn.

McDaniel, C.N. and J.M. Gowdy. 2000. *Paradise for Sale. A Parable of Nature*. Berkeley: University of California Press.

Müller, A. 1909. 'Nauru', *Medizinal-Berichte über die Deutschen Schutzgebiete für das Jahr 1907/08*, 482–496.

——— 1910a. 'Nauru', *Medizinal-Berichte über die Deutschen Schutzgebiete für das Jahr 1908/09*, 452–463.

——— 1910b. 'Eine epidemisch auftretende Erkrankung des Nervensystems auf Nauru', *Archiv für Schiffs- und Tropen-Hygiene* 14, 535–543.

——— 1911. 'Nauru', *Medizinal-Berichte über die Deutschen Schutzgebiete für das Jahr 1909/10*, 546–556.

Nocht, B. 1920. 'Beriberi', in H. Schnee (ed.), *Deutsches Kolonial-Lexikon Band 1*. Leipzig: Quelle & Meyer. Retrieved 20 September 2015 from http://www.ub.bildarchiv-dkg.uni-frankfurt.de/Bildprojekt/Lexikon/Standardframeseite.php

Ratzel, F. 1890. 'Völkerkarte von Ozeanien und Australasien', in F. Ratzel, *Völkerkunde, Zweiter Band, Die Naturvölker Ozeaniens, Amerikas und Asiens*. Leipzig: Verlag des Bibliographischen Instituts, [between pp. 4 and 5].

Senfft, A. 1896. 'Die Insel Nauru', *Mittheilungen von Forschungsreisenden und Gelehrten aus den Deutschen Schutzgebieten* 9: 101–109.

Sonnenschein, F. 1889. 'Aufzeichnungen über die Insel Nauru (Pleasant Island)', *Mittheilungen von Forschungsreisenden und Gelehrten aus den Deutschen Schutzgebieten* 2: 19–26.

Steinbach, E. 1893. 'Bericht über die Gesundheitsverhältnisse des Schutzgebietes der Marshall-Inseln in der Zeit von Januar 1892 bis März 1893', *Mittheilungen von Forschungsreisenden und Gelehrten aus den Deutschen Schutzgebieten* 6: 306–313.

——— 1895a. 'Bericht über die Gesundheitsverhältnisse der Eingeborenen der Marshall-Inseln im Jahre 1893/94 und Bemerkung über Fischgift', *Mittheilungen von Forschungsreisenden und Gelehrten aus den Deutschen Schutzgebieten* 8: 157–171.

——— 1895b. 'Die Marshall-Inseln und ihre Bewohner', *Verhandlungen der Gesellschaft für Erdkunde zu Berlin* 22: 449–488.

——— 1896. 'Einige Schädel von der Insel Nauru (Pleasant Island)', *Verhandlungen der Berliner Gesellschaft für Anthropologie, Ethnologie und Urgeschichte* 28: 545–551.

Steudel, E. 1909. 'Der ärztliche Dienst in den deutschen Schutzgebieten', *Deutsches Kolonialblatt* 20: 921–926.

Thurnham, D.I. 2005. 'Beriberi', in M.J. Sadler, J.J. Strain and B. Cabarello (eds), *Encyclopaedia of Human Nutrition*, 2nd edn. New York: Elsevier, pp. 269–278.

Verhandlungen des Reichstages. 1907a. 'Marshall-Inseln. Missionsberichte'. Denkschrift über die Entwickelung der deutschen Schutzgebiete in Afrika und in der Südsee. Berichtsjahr 1. April 1905 bis 31. März 1906. Aktenstück zu Nr. 41. *Verhandlungen des Reichstages. 12. Legislaturperiode, 1. Session. Bd. 239. Anlagen zu den Stenographischen Berichten*. Berlin: Norddeutsche Buchdruckerei.

——— 1907b. 'Marshall-Inseln'. Denkschrift über die Entwickelung der deutschen Schutzgebiete in Afrika und in der Südsee. Berichtsjahr 1. April 1905 bis 31. März 1906. Aktenstück zu Nr. 41. *Verhandlungen des Reichstages. 12. Legislaturperiode, 1. Session. Bd. 239. Anlagen zu den Stenographischen Berichten*. Berlin: Norddeutsche Buchdruckerei.

——— 1907c. 'Marshall-Inseln. Bevölkerungsstatistik'. Anlagen zur Denkschrift über die Entwickelung der deutschen Schutzgebiete in Afrika und in der Südsee. Berichtsjahr 1. April 1905 bis 31. März 1906. Aktenstück zu Nr. 41. *Verhandlungen des Reichstages 12. Legislaturperiode, 1. Session. Bd. 239. Anlagen zu den Stenographischen Berichten*. Berlin: Norddeutsche Buchdruckerei.

——— 1907d. 'Marshall-Inseln. Klima und Gesundheitsverhältnisse'. Denkschrift über die Entwickelung der deutschen Schutzgebiete in Afrika und in der Südsee. Berichtsjahr 1. April 1905 bis 31. März 1906. Aktenstück zu Nr. 41. *Verhandlungen des Reichstages. 12. Legislaturperiode, 1.Session. Bd. 239. Anlagen zu den Stenographischen Berichten*. Berlin: Norddeutsche Buchdruckerei.

——— 1908. 'Marschall-Inseln'. Anlagen [zur Denkschrift über die Entwickelung der deutschen Schutzgebiete in Afrika und der Südsee im Jahre 1906/07]. Aktenstück zu Nr. 622. *Verhandlungen des Reichstages 12. Legislaturperiode, 1. Session. Bd. 245. Anlagen zu den Stenographischen Berichten*. Berlin: Norddeutsche Buchdruckerei.

——— 1909a. 'Marschall-Inseln. Klima und Gesundheitsverhältnisse'. Anlagen [zur Denkschrift über die Entwickelung der deutschen Schutzgebiete in Afrika und der Südsee im Jahre 1907/08]. Aktenstück zu Nr. 1106. *Verhandlungen des Reichstages 12. Legislaturperiode, 1. Session. Bd. 252. Anlagen zu den Stenographischen Berichten*. Berlin: Norddeutsche Buchdruckerei.

——— 1909b. 'Jahresbericht der Mission vom heiligsten Herzen Jesu in den Marschallinseln vom 1. April 1907'. Anlagen [zur Denkschrift über die Entwickelung der deutschen Schutzgebiete in Afrika und der Südsee im Jahre 1907/08]. Aktenstück zu Nr. 1106. *Verhandlungen des Reichstages 12. Legislaturperiode, 1. Session. Bd. 252. Anlagen zu den Stenographischen Berichten*. Berlin: Norddeutsche Buchdruckerei.

——— 1909c. 'Marschall-Inseln'. Anlagen [zur Denkschrift über die Entwickelung der deutschen Schutzgebiete in Afrika und der Südsee im Jahre 1907/08]. Aktenstück zu Nr. 1106. *Verhandlungen des Reichstages 12. Legislaturperiode, 1. Session. Bd. 252. Anlagen zu den Stenographischen Berichten*. Berlin: Norddeutsche Buchdruckerei.

——— 1911. 'Marschall-Inseln'. Anlagen [zur Denkschrift über die Entwickelung der deutschen Schutzgebiete in Afrika und der Südsee im Jahre 1908/09]. Aktenstück zu Nr. 179. *Verhandlungen des Reichstages 12. Legislaturperiode, 2. Session. Bd. 271. Anlagen zu den Stenographischen Berichten*. Berlin: Norddeutsche Buchdruckerei. The *Verhandlungen des Reichstages* are available online. All retrieved 20 September 2015 from http://www.reichstagsprotokolle.de.

Walbaum, H. 1908. 'Nauru', *Medizinal-Berichte über die Deutschen Schutzgebiete für das Jahr 1906/08*: 230–262.

Wedgwood, C.H. 1936. 'Report on Research in Nauru Island, Central Pacific', *Oceania* 6(4): 359–391.
Wiener, M. 1990. *Ikonographie des Wilden*. Munich: Trickster-Verlag.

3
The War on the *Anopheles* Mosquito
Malaria, Labour and Race in the New Hebrides, 1925–1945

Jean Mitchell

> ...owing to the need for labour of those accustomed to the tropics, it has always been in the interests of the settlers that the native population shall be alive and healthy.
>
> —W.H.R. Rivers, 'The Psychological Factor'

> The 'mosquito experts', called entomologists and the engineers, control malaria by attacking and outwitting the mosquito.
>
> —New Hebrides British Service, 'Notice'

In 1925 a medical entomologist, Dr Patrick A. Buxton, from the London School of Hygiene and Tropical Medicine, embarked on an expedition to study disease in New Hebrides (now independent Vanuatu). He recorded thirteen species of mosquitoes (Perry 1946: 9) and implicated one species, the *Anopheles farauti*, in the transmission of both malaria and filariasis. His research was one of the first scientific exercises to focus on malaria in the southwest Pacific archipelago jointly administered between 1906 and 1980 by France and Britain in an arrangement called the Condominium.[1] Buxton had been researching insects in Samoa before he travelled to New Hebrides and had already conducted a study on malaria in Palestine. He would later undertake research on the tsetse fly in Africa and conduct experimental trials on the use of DDT in London during the Second World War. In New Hebrides, Buxton was accompanied by Dr Sylvester M. Lambert, an expert in tropical medicine, who was working in the Pacific region under the auspices of the Rockefeller Foundation, the

American philanthropic institute established in 1913, which took up the 'mantle of global control, or even, eradication of hookworm, yellow fever and malaria' (Stapleton 2005: 517).[2] The tropical medical experts who undertook their four-month research trip throughout the archipelago on the British government yacht Euphrosyne were encouraged by Condominium officials and were assisted by two unnamed prisoners from Tanna as well as Malakai Veisamasama, a medical practitioner from the British colony of Fiji (Buxton 1926: 421; Widmer 2013; Bennett 2014). Buxton's research was significant because he employed the sciences of entomology and tropical medicine to understand diseases such as malaria and vectors such as mosquitoes about which little was known. He linked malaria to the problem of persistent population decline in New Hebrides, a trend that concerned colonial officials.

Less than twenty years later, New Hebrides became an unintended laboratory for malaria research when American troops landed there to halt Japanese expansion southward in the Pacific during the Second World War.[3] The American military was ready to mount a war effort but they were unprepared for the assault of malaria. The forces were incapacitated at the same time that quinine supplies, the only effective suppressant, had been cut off by the Japanese forces. Historian Leo Slater described the subsequent war time search for malaria chemotherapy as, arguably, 'the largest biomedical research effort of the first part of the twentieth century' (2004: 108), becoming the formative 'model for later biomedical and health research programmes' (ibid.: 109). Controlling malaria was the focus of international scientific research and local disciplinary measures that were instrumental in winning the war. In 1944 General MacArthur, Commander of the Allied Forces in the Pacific Theatre, identified the progress in the war on the mosquito as 'one of the greatest victories' in the southwest Pacific area. It was, in his words, 'a victory by science and discipline over the *Anopheles* mosquito' (cited in Russell 2001: 117). The American military's war on the *Anopheles* mosquito and malaria parasite introduced Melanesians to the new disciplines of scientific and biomedical technologies (Lindstrom 1998: 407; Bennett 2006: 48).

In this chapter I explore the ways in which malaria research undertaken by Buxton in the 1920s and that of American scientists in the Second World War presumed and reproduced ideas of human difference through shifting notions of race in scientific projects that sought to understand and control the *Anopheles* mosquito and its environment in New Hebrides. By examining how the work of Buxton and the American entomologists proceeded through localized attention on mosquitoes, New Hebridean Islanders and indentured Tonkinese labourers, I shall investigate what Alexandra Widmer has referred to as 'the malleability of racial thinking and the intransigence of racial categorization' (2012: 69). This chapter tracks 'how science travels in the modern world and what it does when it arrives' (Anderson 2008: 7). Following the work of entomologists in New Hebrides also points to the ways in which 'global science is assembled out of a set of rather peculiar local achievements', as Warwick Anderson further contends (ibid.). Attending to these local achievements in New Hebrides in the context of malaria research provokes a number of questions. How is 'racial thinking' reproduced through the science of disease control in colonial and military medicine? How did Buxton's research inform his thinking about difference and further his understanding of depopulation and malaria? What happens when endemic malaria becomes an epidemic among soldiers in an occupying army during war? Finally, how did war-time deployments of science and discipline to combat malaria's vector, the *Anopheles* mosquito, shape difference and the ways in which the disease and race were constituted or reconstituted?

Malaria, endemic in many parts of the tropics (Packard 2007), is now understood as a complex disease configured through the human host, the *Anopheles* mosquito, the *plasmodium* parasite and their environments. Race had long been implicated in malaria transmission and control in colonial projects such as those of New Hebrides (Bennett 2006), India (Arnold 1999) and the Philippines (Anderson 2006). David Arnold, in his research on malaria in India, argues that locating race alongside debates about science, disease and the environment demonstrates the shifting and multiple deployments of racial categories and understandings. In New Hebrides, such debates were centred on the issue of depopulation that invariably drew the attention of researchers there before the war (Rivers 1922; Baker 1928; Harrison 1936). This problem also preoccupied colonial authorities but their concern receded in colonial records after the Second World War (Widmer 2013). Colonial responses to malaria varied, but as Megan Vaughn has argued from the vantage point of Africa, 'the power of colonial medicine lay not so much in its direct effects on the bodies of its subjects ... but in its ability to provide "naturalized" and "pathologized" accounts of those subjects' (1991: 25).

While malaria had long troubled colonial expansion, scientific understanding of this ancient and widely dispersed scourge was slow and multi-sited. It was only in the latter half of the nineteenth century that scientists developed germ theory and understood that infections are caused by minute organisms. Alphonse Laveran, a French army officer working in Algeria in 1880, identified *plasmodium*, the pathogenic agent that causes malaria. In 1898 Ronald Ross, a British medical officer in India, following the research leads of his mentor, Patrick Manson, discovered the transmission of malaria to birds by mosquitoes. Ross demonstrated that the infective stage of the malaria parasite was injected into the host when the mosquito released saliva into the bite prior to ingesting blood (Patterson 2009: 189; Cirillo 2011: 381). In 1898 Italian Giovanni Grassi linked Anopheline mosquitoes to the transmission of malaria to humans (Packard 2007: 115). With new theories and technologies such as the achromatic microscope in the 1880s (Packard 2007: 116; Cirillo 2011), the quest to identify and control pathogens intensified as colonialism 'set into motion the spread of public health administrators that saw the colonies as their own laboratories in which to develop Western Science' (Mitman, Murphy and Sellers 2004: 12).

Ross, who won a Nobel Prize for his work in malaria transmission, became a leading expert on malaria and its eradication. He believed that malaria impeded the extension of 'civilization' and the productivity of labour in all corners of the colonized world (Ross 1923: 15). The adverse effects of malaria on colonial settlers and administrators often referred to as the 'white' populations and the need to secure labour 'accustomed to the tropics' (Rivers 1922: 89) were common concerns in many different colonial projects (Arnold 1999: 128). From the 1880s, securing labour for the British and French plantation was a persistent theme. During the Second World War the American military also desperately required labourers. The demand for indigenous labourers and the adverse effects of malaria were further complicated by the phenomenon of acquired partial immunity in areas of endemic malaria that rendered New Hebrideans 'carriers' of the disease.

The work of Buxton and Lambert in New Hebrides and the organizations they represented marked an important turning point in the inter-war period as colonial governments attempted to study and control diseases through the establishment of new schools of tropical medicine, inter-governmental health organizations and the League of Nations (Packard 2007: 115). The fledgling ecological approaches directed to vector-transmitted diseases were significant for they instituted strategies such as those undertaken by the Rockefeller Foundation that displaced earlier approaches to disease that focused on race (Anderson 2006: 209). Drawing on his research in the Philippines, Anderson argues, however, that such ecological and technical approaches failed to eradicate racialist assumptions before the Second World War (ibid.). Alexandra Widmer's work on how 'racial knowledge was scientifically produced' (2012: 72) through encounters between researchers and the inhabitants of New Hebrides underlines the importance of examining data production and interpretation as well as programmes that reinforce hierarchies of value (Dureau and Low 1999). I shall argue that Buxton's research on the *Anopheles* mosquito and malaria in New Hebrides did not succeed in decentring race. I shall further argue that the focus on science to combat the *Anopheles* mosquito in New Hebrides during the Second World War both privileged environmental control and racialized the disease in new ways. The American war effort to contain the New Hebridean malaria 'hosts', to eradicate the vector and to treat malaria parasites served to show the complexity and the specificity of these 'local' relationships through science. At the same time the well-being of the population, whose characterization moved from vulnerable in the context of pre-war depopulation debates to dangerous in war time, was considered a by-product of the science and discipline required for war. Malaria and the enactments of race, I shall contend, are amenable to a host of signifying practices that locate and naturalize differences in new and old ways.

Depopulation and Malaria

Ideas about racial difference in New Hebrides coalesced in debates about depopulation (Widmer 2012, 2014). The decline in population, in Buxton's view, 'almost warrants one in saying that the prediction of the eventual extinction of many of the races now living in New Hebrides is inevitable' (1926: 453). The preoccupation with population decline drew attention to the impact of epidemics and the low rates of fertility throughout the archipelago. Buxton noted that at the time of his visit in the mid-1920s little was known about the diseases of New Hebrides and many of the explanations of depopulation were 'erroneous' (ibid.: 423). The labour trade within the Pacific and the epidemics that began with contact, he argued, have contributed to depopulation in New Hebrides but could not fully explain why the population did not rebuild (ibid.: 441). The endemic diseases of malaria and yaws, according to Buxton, contributed greatly to the persistent problem of depopulation (ibid.) in New Hebrides where malaria occurred on all the islands except Futuna. In Buxton's view, 'the severity of malaria on all of the islands is one of the factors which determines the power of a population to recover from epidemics and from the labour trade' (ibid.: 427).

The scientific narratives that circulated in the first part of the twentieth century linked depopulation to differences that were cultural, biological and psychological. Melanesians were regarded as a 'race in decline' due to 'the lack of biological resistance or the inability or unwillingness to adapt' (Widmer 2012: 82–84). W.H.R. Rivers, a medical doctor and anthropologist from the University of Cambridge, made two research trips to New Hebrides. He recognized that introduced diseases such as tuberculosis were still problematic, but he believed that 'the people are largely habituated to the malaria which certainly existed among them before the coming of European influence' (1922: 89–90). John R. Baker, a zoologist from Oxford University, discussed depopulation on the Island of Espiritu Santo based on several research trips to New Hebrides during the inter-war period. Population decline, argued Barker, was due to a combination of factors that included diseases brought by Europeans as well as injurious local and indigenous practices. Malaria, in his view, was not a primary cause but together with yaws probably contributed to lowered resistance (1928: 295). Baker sought to improve scientific method by taking 'vital statistics of a large part of Espiritu Santo and as far as possible to find whether a suggested cause was operative or not by reference to those statistics' (ibid.: 333).

The Buxton Line: Demarcating Malaria and Difference

One of Buxton's scientific achievements, as noted, was the successful identification of the *Anopheles farauti* as the vector of both malaria and filariasis.[4] It was significant as measures to control this particular mosquito could decrease the incidence of both endemic diseases. While early entomologists tended to ignore protozoa in order to concentrate on the vector and various species of mosquitoes (White 1995: 233), Buxton also attempted to establish the spleen and parasite rates by examining the spleens of 280 children and taking the blood-film data of 209 children. However, the sample was too small to enable generalizations (Buxton 1926: 433) and he attempted to support his research by generating data from church records. Buxton registered his frustration in securing data from unwilling and 'heathen' people who could not distinguish the categories of child and adult (Widmer 2012: 78–79). To complicate matters, Buxton's research took place over an unusually dry season allowing him to take only five samples of mosquito larvae in Santo, one of the most malarious islands (Daggy 1945: 1). American entomologists later produced large amounts of data on the taxonomy, habitat and behaviour of the vector replete with statistics and drawings (Belkin, Knight and Rozeboom 1945; Daggy 1945; Perry 1945a).

While his entomological work was thwarted by the dry season, Buxton left his mark by designating the Buxton Line that demarcated the range of the *Anopheles* mosquito. Southern Vanuatu was the eastern limit of the *Anopheles* mosquito in the Pacific where the fauna of the Pacific spreads out from the west, becoming thinner in the east as does the population of the *Anopheles* vector (Buxton 1927; Iyengar 1955: 1; Opeskin 2009: 17). The Buxton Line inscribes Melanesia (with the exception of New Caledonia) as the malarious region of the Pacific, underlining endemic malaria as an essential difference between Melanesia and the rest of the Pacific region. Polynesians did not have malaria nor did the Polynesians living on the island of Futuna in New Hebrides. In the 1830s French navigator and naturalist, Dumont d'Urville, classified the South Pacific population into three distinct groups, Polynesians, Melanesians and Micronesians. His categories were racialized 'based on skin colour, physical appearance, language, political institutions, religion and reception of Europeans' (Douglas 2008 : 9). Dumont, as historian Bronwen Douglas argues, sharply differentiated the Melanesian or 'black oceanian race' as inferior to the 'copper-coloured race' of Polynesians (ibid.: 26). Scholars have demonstrated that since the earliest accounts of Europeans, Melanesians have been depicted as less 'civilized' than Polynesians (Thomas 1994: 101). While the prevalence of malaria signified the vulnerability of Melanesian populations, it also indexed their primitivism. The Buxton Line was used to encompass social and perceived biological differences. Drawing on his research in Samoa, Buxton noted in New Hebrides, 'The natives are in a much more primitive state than the Samoans' and in contrast to their Polynesian neighbours they are 'still good bushmen' (1927: 424). Malaria was another means to draw a 'line' between civilization and primitivism in the South Pacific (ibid.: 455; Ross 1923).

Buxton's description of New Hebrides elided the past and present allowing him to make authoritative descriptions of Melanesians that subtly linked them to malaria and their propensity for population decline. He articulated differences that distinguished the physical, social and spatial features of New Hebrideans. According to Buxton, the people of New Hebrides had no concept of 'natural causes' in the case of sickness and death that were usually attributed to sorcery (1927: 423). This contention was undermined by anthropological descriptions of New Hebrideans and their use of the medical services provided by missionaries in some areas of the archipelago. Local inhabitants were selective in their choices and weighed the efficacy of the medical service on offer (Widmer 2012: 79–82). Rivers also indicated a far greater complexity of local conceptions of morbidity and mortality by suggesting that 'native therapeutic ideas were not adapted to the new diseases' (1922: 90). The large-scale programme of injections aimed at eradicating yaws that Lambert initiated in the 1920s and 1930s was readily accepted (Laing 1989). Robert Black, an Australian expert in tropical medicine who conducted research on malaria during the Second World War, also noted that people 'soon learned that quinine was a potent counter magic in the case of malaria' (1956: 141). While identifying malaria as problematic, Buxton simplified the worldview of the inhabitants and underestimated their capacity to comprehend sickness and their efforts to protect their health. This distrust of the 'native's' ability to act in the arena of health, evident in Buxton's contention, facilitated increasing attention to vectors such as mosquitoes that were seen to be more amenable to control than was the 'native', as Packard has convincingly argued (2007: 118).

Ideas about biological and cultural differences were integral to researchers' understanding of the relationship between quarantine and malaria. John R. Baker promoted the idea of quarantine by advocating an 'experiment' to isolate the inhabitants on the island of Gaua in the northern region of the archipelago in order to gain insight into population decline and to facilitate recovery of the population (1928: 299). Buxton construed the spatial and social organization of villages in New Hebrides as an effective quarantine against disease in general and malaria in particular. Fear of magic and inter-group hostility, according to Buxton, limited interaction between neighbouring villages, creating a 'system' that 'amounted to a permanent strict quarantine between one community and the next' (1927: 423). This depiction of the social relations that existed among villages was at odds with those of social anthropologists who had already described social relations in considerable detail (Rivers 1922; Speiser 1922). Drawing on his interest in intersex pigs, Baker disagreed with Buxton's claim that Islanders were isolated from each other for, he argued, there were active exchanges of pigs (Baker 1929a). Buxton is one of many observers (Black 1956: 137) to suggest that contact with Europeans and labourers' movements between villages and plantations ruptured this quarantine (Buxton 1927: 427), thereby endangering the health of New Hebrideans. While quarantine is an important factor in the context of epidemics, Buxton's reliance on its explanatory value undermined the complexity of interactions among Islanders and between Europeans and Islanders (Widmer 2008: 103). With the exception of malaria, Islanders did not have the infectious diseases that afflicted Europe, Africa and Asia, nor did they have immunity to these diseases, which resulted in epidemics on contact with Europeans (Shlomowitz 1992: 116). Malaria, which was endemic, was intensified by the introduction of the plantation economy and the resettlement of villagers to coastal areas. Islanders who were resettled to coastal villages by missions and colonial governments and those plantation labourers who worked in poor conditions in lowland areas were more susceptible to malaria transmitted by the coastal-dwelling *Anopheles farauti* mosquito (Shlomowitz 1992: 117; Denoon 1999: 329). These developments, integral to the colonizing projects of both the British and French, were obscured in Buxton's explanation of indigenous quarantine.

Labour and the Cultivation of Difference

The adverse effects of malaria on the European population and the need to secure labour were colonial concerns that shaped measures to extract labour and to protect the colonial settler population. Protective practices underlined the necessity of containing the indigenous populations, curtailing their movements, living separately from them to prevent contagion and, in some places, treating the New Hebrideans who were designated as the carriers of the *plasmodium* parasite. Describing malaria in the American-occupied Philippines before the First World War, Anderson notes that it was believed that 'the greatest source of danger to the white man in a malaria locality lies in the native population, especially in the native children' (2006: 208).[5] Therefore, he continues, it was considered 'futile to rid any locality of malaria as long as the native element in the question is neglected' (ibid.: 209). In New Hebrides missionaries, traders, planters and officials suffered from malaria (Buxton 1927: 435; Thompson 1981) and according to anthropologist Felix Speiser, 'malaria does incomparably more harm to the white than the native' (1922: 28; see also Baker 1935). In the late 1880s, efforts to establish plantations were 'defeated by a combination of native hostility and malaria' (Thompson 1981: 35). However, in 1888 a French Naval medical officer brought quinine to New Hebrides, possibly offsetting some of the lethal effects of the disease among the French population (Laing 1989). The Anglo-French Condominium administration did little to develop the health care infrastructure and largely left the provision of medical services to the missionaries. Presbyterians operated clinics and small hospitals from the late nineteenth century. Lambert's efforts from the 1920s represented the first attempts to extend government-funded biomedical health programmes in New Hebrides with professionally trained Pacific Islanders (Widmer 2013). While few resources were allocated to protecting the health of the inhabitants of New Hebrides, they were routinely exposed to conditions such as relocation, overwork and poor nutrition that made them vulnerable to malaria. Thompson has argued that from the 1880s there was increased pressure on declining 'native' populations to work on French plantations. The shortage of labour resulted in unethical recruitment and poor treatment of labourers (1981: 9). Speiser, in fact, identified long-term labour contracts on plantations within New Hebrides as obstacles to population recovery (1922: 57–58). The link between the labour conditions and malaria was recognized at the turn of the century by a number of malariologists who identified social and economic forces as critical factors in controlling malaria. Increasingly, however, a preoccupation with the biometrics and biomedicine of malaria evident in the war-time strategies in New Hebrides bypassed approaches, which emphasized the political economy of the disease (Packard 2007: 111–112).

The search for labour, as noted, was a persistent theme within the Pacific region. French and British planters in New Hebrides seeking local labourers faced competition from the Pacific labour trade in the region that operated from the 1860s into the early twentieth century and later competition from the nickel mines and domestic service in New Caledonia, a nearby French colony (Shineberg 1999; Bennett 2014). However, with such competition for their labour as well as access to their own land, New Hebrideans could be neither coaxed nor coerced into working on long-term contracts on plantations in New Hebrides, favouring instead causal over contracted labour. The severe shortage of labour was resolved for the French by the introduction of indentured labour from Tonkin (Adams 1990: 140). The Tonkinese were a significant presence by 1925 (MacClancy 1981: 89) and comprised 10 per cent of the total population in New Hebrides by 1929 (Meyerhoff 2002: 47). The arrival of the Tonkinese indentured labourers presented a new configuration of difference and pathology in the context of the local and European populations. Baker, for example, suggested they 'are probably responsible for certain epidemics which have not been identified' (1928: 287). Buxton recommended that the Tonkinese be strenuously screened for tuberculosis, syphilis and intestinal parasites but there was no mention of malaria. He advocated, however, that the right to secure foreign indentured labour be extended to the British planters in order to protect the fragile indigenous population. Insisting that the indentured workers be segregated to avoid racial mixing or 'hybridization' (1927: 430), Buxton categorized the Tonkinese as the 'imported race' which could be a menace to the 'the natives' and to 'White Australia' if they were not repatriated at the end of their contracts (ibid.). Tonkinese labour was a temporary means to insulate New Hebrideans from the excesses of plantation labour as Nicholas Thomas has argued in the Fijian context of indentured labourers (1990: 158).

Linking malaria to persistent depopulation, Buxton advocated the protection of the vulnerable indigenous population through such measures as separation from 'imported races' and the extension of the use of indentured labourers. In the 1920s, Buxton's medical entomological research contributed to scientific knowledge about *Anopheles* mosquitoes and malaria in New Hebrides at a time when little research had been undertaken. However, with the advent of the Second World War and the arrival of American forces, the science of entomology would be deployed in new ways in New Hebrides, generating extraordinary data on mosquitoes in order to control malaria (Spencer 1992: 33) that not only threatened military personnel but the outcome of the war. The problem of the fragility of the 'native' population disappeared from view as they became regarded not as necessary labour pools but as carriers of malaria and therefore potentially dangerous to the American forces. Antje Kühnast in this volume also demonstrates how colonial depictions of Nauruans changed to fit the contours of colonial and extractive economic imperatives. Population decline and the failure to survive epidemics in Nauru were linked to racial traits and propensities.

War and the Science of Malaria Control

The Second World War transformed 'the character of science in a fundamental and irreversible way', unleashing 'an epistemological revolution' (Fortun 1993: 596) and changing research models and the relationships among military, civilian and scientific institutions. According to Roy MacLeod, it was the sheer size of the American effort in the Pacific and the commitment to 'bring science into field conditions' (1999: 5) that defined the science of the Pacific War. The war also changed the trajectory of malaria control by deploying entomologists and engineers to combat the disease in the Pacific. In 1942 malaria posed a more serious threat to the marines who landed in New Hebrides than did the advancing Japanese army. Throughout Melanesia, malaria quickly reached epidemic proportions among military forces who had no immunity to the disease.[6] In New Hebrides 'outbreaks of malaria reached crippling, epidemic proportions among navy and marine corps forces' (Beadle and Hoffman 1993: 323). On the island of Efate where the capital Port Vila is located, 'the primary attack rate of malaria (or the rate of original admissions for malaria) peaked at 2,632 cases per 1,000 average strength year for all US and Allied forces in April, 1942' (ibid.). The problem was compounded on Efate by the absence of suppressive drugs such as quinine. Although the incidence of new malaria cases dropped sharply in the last part of 1943, readmissions continued to be high (ibid.).

In April 1942, Navy entomologists were sent to support the battalion that occupied Efate, where malaria was 'devastating' the marines. A few months later, Ensign Kenneth L. Knight, the first Navy entomologist to work in a combat zone, was deployed to Efate in New Hebrides as part of the Navy Epidemiology Units (NEU). By 1944 there were 122 epidemiology units and by the end of the war there were approximately 150 units with 900 personnel and 200 entomologists working in the southwest Pacific (Gerber 2008). The dramatic decline of the malaria case rate among the military by 1944 was hailed as 'an achievement of historical importance in preventive medicine' (Link and Coleman 1955: 807). American researchers collected and collated large amounts of entomological, epidemiological and parasitological data through the mobile entomological teams and the Malaria Control and Survey Units (Spencer 1992: 33).

During the war a military apparatus was erected on the islands of Espiritu Santo and Efate at a time when the population of New Hebrides was widely believed to be in a vulnerable state. On the island of Espiritu Santo alone, there were about 100,000 American and allied troops stationed during the course of the war (Laing 1989: 145). The war effort in New Hebrides required local labourers in order to free troops for combat (Beadle and Hoffman 1993: 331). Employment with the military, despite its long hours and heavy workloads, served Islanders' interest in short-term contracts, offered better pay than plantations and provided access to food and other commodities (Lindstrom 1998: 407). However, as malaria erupted among American troops, the local labourers were identified as 'reservoirs' of infection (Beadle and Hoffman 1993: 332). When the research of American entomologists confirmed the danger of proximity to 'native' labourers (Daggy 1945: 7), medical surveillance and treatment of labourers and villagers near the bases became an essential part of the malaria control programme aimed at curbing the disease and winning the war.

The high rates of malaria were attributed to 'mistakes' made in the early stages of the southwest Pacific campaign. One such mistake was linked to the military's failure to estimate the impact of proximity to the infected local inhabitants who were employed to do manual labour (Beadle and Hoffman 1993: 332). As noted, this allowed close contact with infected *Anopheles* mosquitoes and the transmission of the disease to the susceptible American troops (ibid.: 331; Bennett 2006: 40). The problem was that the local labourers and villagers living near the American bases had, as Rivers noted decades earlier, become 'habituated' to the disease. Through the rigours of repeated infections in endemic areas Islanders had acquired a degree of immunity to malaria making it possible for them to live in malarious areas that could kill a non-immune adult (Black 1956: 136). Robert Black explains: 'In an undisturbed community malaria tends to exist at a certain level or degree of endemicity. Endemic malaria causes acute and chronic illness and a high infantile mortality rate. Those adults who survive acquire a varying degree of tolerance to the malaria parasites occurring in their own areas' (ibid.; Bennett 2006: 29; Webb Jr. 2009: 157). Immunity within the population of New Hebrides also varied as the *Anopheles faruati* was primarily a coastal dweller (Daggy 1945). Harrison in the 1930s had noted that people living in hill areas and inland had little or no mosquito-borne disease (1951: 338). Those who lived above 1500 metres had less immunity than coastal people for if the intermediate populations were sparse, people had less exposure to the parasites that were necessary for immune responses (Bennett 2006: 29). While labourers who hosted malaria parasites were problematic for the Americans without immunity, the transmission of malaria was not one-way. American military returning to New Hebrides from malarious areas such as Guadalcanal, Solomon Islands infected local inhabitants and other Americans (Beadle and Hoffman 1993: 332; Bennett 2006: 43). Almost every man who served on Guadalcanal between August 1942 and February 1943 'fell victim to the disease' and these men then 'acted as their own reservoirs of malaria when there were no infected local inhabitants' (Beadle and Hoffman 1993: 332).

The War on the *Anopheles* Mosquito

In order to fight malaria each Marine Division was provided with a malaria control team (called an Epidemiology Unit) comprised of three officers (a malariologist, an entomologist and a parasitologist), twelve enlisted men and labourers (Gerber 2008: 23f.). Survey and Control Units were established in New Hebrides for 'the sole purpose of finding and eliminating mosquitoes in as short a time as possible' (Perry 1945a: 8). With this clear mission the units worked in tandem to research and to apply their findings in order to combat the *Anopheles* mosquito. The Survey Unit determined the malaria species present, defined the predominant vectors, assessed the malaria rates among 'native' populations and did troop and community education (Joy 1999: 202). Surveys of both the mosquito and New Hebridean populations were considered necessary and were ongoing. The Control Unit undertook measures to decrease the vector to human interaction through the eradication of the breeding sites of mosquitoes, aerial spraying of adult mosquitoes, screening buildings and training labourers (Beadle and Hoffman 1993: 322; Joy 1999: 202). Reducing human-mosquito transmission focused attention on local labourers and involved actions such as transferring malarious night workers to daytime shifts and removing native residents away from military installations (Beadle and Hoffman 1993: 322).

Controlling malaria was challenging for, in contrast, to many other species, the *Anopheles farauti* showed an 'exceedingly wide range in its adaptation to various types of aquatic habitats' (Daggy 1945: 6). It readily utilized temporary man-made breeding places as well as permanent breeding sites such as rivers, streams, taro gardens, lakes, swamps, marshes and open wells (ibid.: 2). The vector was found to be more abundant in inhabited areas where roads, villages, gardens and wells existed (ibid.: 5). The entomologists found, however, that the *Anopheles* mosquito did not breed in the water collected in coconut shells left on the ground (Perry 1945b: 4).

This was clearly a common perception as the colonial administration had in 1942 issued a regulation for malaria control ordering the collection of empty coconut shells on cultivated land (Widmer 2008: 104; Opeskin 2009: 189). The research findings on the extensive temporary breeding sites common in the rainy season and the range of its permanent breeding sites far exceeded Buxton's observations of breeding sites (Buxton 1927: 435). The opportunistic breeding sites, according to the American entomologists, 'constitute the major problems in anopheline control' (Belkin, Knight and Rozeboom 1945: 241; Daggy 1945: 5). Prior to the war there had been no research available on the life cycle and the flight ranges of the *Anopheles* mosquito in New Hebrides. During the war, it was found that mosquitoes in Espiritu Santo usually flew 400 yards from their breeding sites but they could fly up to a mile (Perry 1945b: 9; Daggy 1945: 15). The mosquitoes were also able to fly 600 yards across open water 'to attack' the military aboard anchored ships (Belkin, Knight and Rozeboom 1945: 261; Daggy 1945: 8, 13). Data on the flight ranges were significant for malaria control as they indicated the distance required to protect military personnel from 'native huts' (Daggy 1945: 11) that were often found to be the diurnal resting places of the female *Anopheles*. Research concurred that the 'blooded' female inhabited dark places such as 'occupied houses and native's huts' and that females live up to thirty-five days (Belkin, Knight and Rozeboom 1945: 261). Intensive mosquito control measures operationalized within the flight radius of the Anopheles mosquito greatly reduced the malaria rate (Daggy 1945: 11).

The entomologists working in the field and in laboratories kept detailed, daily records over a period of approximately two years in coastal areas. They collected and reared adults from individual larvae taken from as many natural breeding places as possible. They also studied 'the progeny of isolated females' which had been captured in natural resting places and native dwellings (ibid.: 241; Perry 1945a: 17). The researchers found the *Anopheles farauti* to be 'an efficient vector' with particular and dangerous attributes. 'Its preference for human blood, painless bite, lack of hum and habit of resting in native huts and shelters' (Daggy 1945: 12) all contributed to its deadly efficacy. The entomologists described the feeding females as 'shy, wary' and 'easily frightened' but 'persistent in returning to attack' (Perry 1945b: 7). The mosquitoes also exhibited a 'wavy and dancing flight when they are attempting to bite' (ibid.: 9).

The entomologists relied on mosquito counts to demonstrate their progress in lowering the density of *Anopheles farauti* (Patterson 2009: 156). Mosquito surveys were regularly made by entomologists and crews under the direction of the Survey Unit in order to determine the size of the mosquito population in a given area. These surveys also identified the different types of mosquitoes and the methods of eliminating them (New Hebrides British Service 1944a: 2). Mosquitoes were collected and counted in the living spaces of villagers and labourers in villages and on plantations. Aerial photos of houses and villages were taken and the interiors of their houses were sprayed with insecticides. The living spaces and the houses of the local people were incorporated into the entomological research. Entomologists explained that mosquitoes:

> are commonly found in their thatched, split bamboo huts between the levels of two and seven feet. In numerous collections … resting adults have on occasion been collected during the daylight hours on the stones which surround the central fireplace and on the undersides of the crude bamboo beds used by the natives. Pots, pans, bottles, jars and other utensils found scattered over the floor of the native huts have been found to be excellent collecting spots for adult anophelines… (Perry 1945b: 6)

Since the 'feeding' Anopheles mosquitoes tended to fly at low levels, frequently biting the extremities of humans, the entomologists also found 'collections are readily made from natives around their ankles and thighs' (ibid.: 7). The bodies or body parts of New Hebrideans were also incorporated into the research on the *Anopheles* mosquito.

Malarial control measures included ambitious efforts to destroy the vector's breeding sites. Engineers, explosives and large machinery were deployed to reshape the environment. The knowledge obtained from research about the life cycle of the mosquito guided the work of the Control Unit. Research showed that anopheline mosquitoes must spend the first seven to ten days in water and thus draining or filling in sources of water was important. It was also found that young mosquitoes can be killed in water by Paris green oil and other chemicals, while adult mosquitoes could be killed by spraying them in their resting places (New Hebrides British Service 1944a: 2). Weeds were cleaned from slow-moving water and larvicides and insecticides were applied with aerial spraying. Aerial spraying of insecticides in areas with a heavy population of mosquitoes was effective. The fish minnow, *Gambusia affinis*, was also introduced to clean up the larvae (Daggy 1945; Joy 1999: 203; Bennett 2006: 35). Many of these measures were not new as the earlier inter-war environmental strategies included measures to eliminate mosquito-breeding sites. However, during the war, the enormous entomological knowledge acquired, the financial and technical resources allocated, the intensity of the effort expended and the discipline with which the vector control programme was enforced were unprecedented and highly effective. The incidence of malaria had already decreased before DDT reached combat theatres in the southwest Pacific in late 1944 (Joy 1999: 202; Patterson 2009: 157).

Man-Made Malaria and Eradication of Mosquitoes

Despite the commitment to malaria control in the early days of the war, the vector's habitat expanded due to the construction of bases and roads undertaken by the Americans. It was estimated that 90 per cent of the temporary breeding places during the rainy season were created by troops in Espiritu Santo during the early occupation of the base (Daggy 1945: 6). As the entomologist Daggy observed, 'some of the permanent breeding places were even improved by American activities' (ibid.). There were concerted efforts to map and control those man-made habitats (Bennett 2006: 35). While the American forces conceded that they created man-made malaria, they were concerned that plantations not only created habitats for the vector but also harboured infected labourers. The Tonkinese cultivation of *Ipomoea aquatic* (water cabbage) to supplement their meagre diets was regarded, for example, as 'intimately associated with mosquito breeding' (Black 1956: 137). In 1942 the Malaria Control Unit appealed to the plantation owners to contribute to the war effort to control malaria by making an argument for the improved productivity of their labour:

> Plantation owners are able to lend valuable aid to mosquito control to protect the health of the Allied Armed forces, and furthering the steps toward Victory. As a matter of economics, when plantation owners protect the health of their labourers, both Tonkinese and native Melanesians, they are able to profit by the increased industry and efficiency of their workers. A sick man, whether he be a soldier or a plantation laborer, can never perform his duties as can a well man. (New Hebrides British Service 1944a: 2)

It was, ultimately, the New Hebrideans who were considered the major source of man-made malaria. Most of the elements of their daily lives were identified as disease producing. Their villages were sited near rivers, they used wells, grew water taro, built terraces and 'primitive irrigation systems', all of which resulted in a heavy *Anopheles farauti* population. According to Daggy, such 'observations only further confirmed the dangers of native villages as centers of malaria infection for troops' (1945: 7). The reduction of the activities of everyday life to sites of man-made malaria and the preoccupation with 'native carriers' of malaria were driven by the singular focus to eradicate the mosquito vector as a necessary condition of military victory.

The American war-time preoccupation with the eradication of the mosquito vector was at odds with the ecological approaches that were developed during the inter-war period. These ecological perspectives – which directly challenged the idea that vector-borne disease could be eradicated (Anderson 2006: 209) – did not survive the inter-war period (Packard 2007: 136). Ecological approaches to diseases, such as malaria, that concentrated on control rather than eradication were predicated on understanding the interaction of the host\vector\parasite triad. Historian Helen Tilley, in her discussion of the tsetse fly in Africa, underlines the complexity of arthropod-transmitted diseases in the inter-war period. She explains that the elements of a control programme that included 'issues of endemicity, social organization, population densities (of both vectors and humans), habitat changes and individual susceptibilities were increasingly brought into a single analytical framework' (2004: 36). Tilley further explains that the behaviour of vectors and disease patterns were increasingly regarded as highly localized and scientists using ecological frameworks built this specificity into control programmes. While American entomologists generated a great deal of data about malaria and its vector, their singular focus on mosquito eradication, the urgency of war and the instrumental approach it dictated undermined any capacity to operationalize a sustainable control programme. Vector-borne disease programmes in New Hebrides during the war were dependent upon the particular science of entomology and a particular production of knowledge (Packard 2007: 118). The focus on the agency of the insect displaced the social agency of New

Hebrideans. Malaria, the *Anopheles* mosquitoes, and the bodies of the Islanders and their living quarters were mapped. As in mapping exercises undertaken in earlier times and places, the New Hebrideans were 'represented as closely tied to nature' (Harrison 2005: 59). Entomological research and malaria eradication programmes spatialized New Hebrideans' social and economic practices in such a way that they were reduced to sites and instances of man-made malaria. The war-time entomologists sought to 'control malaria by attacking and outwitting the mosquito' (New Hebrides British Service 1944a: 2) and in so doing racialized the 'native' body and the danger that it posed. Sarah Ehlers in chapter four demonstrates how colonial encounters with sleeping sickness, a vector-borne disease, resulted in the pathologization of Africans and their bodies, rendering them dangerous and responsible for the 'disorder of the environment' and the presence of the tsetse fly.

The War on The Malaria Parasite

While eradicating the *Anopheles* mosquito preoccupied the entomologists, addressing the effects of the *plasmodium* parasite was also considered vital to ensure combat ready troops in Melanesia. However, during the Second World War there were few drug possibilities as the quinine supply was drastically cut when Japanese forces occupied Dutch East Indies (Indonesia) where 90 per cent of the quinine supply was grown. There was a concerted international effort to find a replacement for quinine, derived from the bark of the cinchona tree that had been used for several centuries to suppress the disease. Over the course of the Second World War 'more than 14,000 compounds were screened for anti-malarial activity' (Slater 2004: 108). The work of Nobel Prize-winning neurologist Camillo Golgi in 1906 deepened understanding of the *plasmodium* parasite by finding the connection between the symptoms of recurrent chills and fever that characterize an attack of malaria and the rupture and release of merozoites into the blood.[7] His research, which explained the efficacy of quinine, facilitated research into finding a replacement. In 1942 clinical research trials of Atebrine, a drug developed in Germany during the First World War, were undertaken to determine the dosage and timing of the use of the drug (Joy 1999; Anderson 2013). The American and Allied forces then decided to use Atebrine and it dramatically reduced troop morbidity, particularly among the American forces (Condon-Rall 1991: 494; Webb Jr 2009: 158).

While Atebrine was found to be highly effective, the American soldiers were often reluctant to take the drug for fear of its rumoured side effects that consisted of yellowing of the skin, suggesting the possibility of liver damage as well as the loss of sexual potency (Walker 1952: 128; Bennett 2006: 35). Discipline was imposed by military commanders to ensure that the soldiers protected themselves from malaria by taking Atebrine and preventing mosquito bites (Joy 1999). The lack of discipline among the soldiers was regarded as another 'mistake' of the early war. Line personnel failed to take adequate preventive measures and before landing in New Hebrides many marines had thrown away their mosquito nets (Beadle and Hoffman 1993: 323; Patterson 2009: 154). The noiselessness and the painless bites of the *Anopheles* mosquito also made disciplined protection among individual soldiers more difficult (Bennett 2006). However, as the incidence of malaria increased, discipline among soldiers became crucial as most of the malaria cases occurred in 1943 when the American forces could not afford to lose soldiers (Beadle and Hoffman 1993: 322). Military officials decided 'Nothing important can be left to the individual inclination of the soldier … specific instructions must be issued and diligently enforced' (Joy 1999: 200). The health of troops and the enforcement of preventive measures to combat malaria became the purview of military leaders (ibid.).

New Hebrideans who lived near the military bases on the islands of Efate and Espiritu Santo and those labourers from various parts of New Hebrides who worked alongside Americans were also disciplined. Their movements were restricted to limit the possibility of infecting American forces with malaria. Their environment was reworked to accommodate the war effort and to eradicate the mosquito vector. They were also brought under medical scrutiny and treatment in particular ways. New Hebrideans were exposed to new medical procedures that were unavailable to most of them before or after the war. Their blood was drawn and screened with microscopes, their spleens were palpated and they were also given anti-malarial drugs such as Atebrine. Lindstrom and Gwero suggest that New Hebrideans often regarded drugs and medical treatments as part of the exchange between them and the American military that included a range of commodities that flooded New Hebrides during the war (1998: 214). Historian Judith Bennett, however, has argued that there was far too little effort made to measure or to appreciate the impact that anti-malarial drugs had on Melanesians who had some acquired immunity and for whom medication would not be available when the war ended (2006: 48). She points out that while the Americans were concerned about the effects of such drugs on their own military forces, this concern did not extend to the local villagers and labourers who were given the same drugs (ibid.: 45). The American war-time effort, according to Slater, while offering a model for future large-scale biomedical research projects, failed to offer adequate care to civilians such as the New Hebrideans (2009: 2). The key problem is that the use of anti-malarial drugs to suppress malaria among labourers and villagers in New Hebrides could have made them unable to deal with new malaria infections and a severe attack of the disease which may occur in the absence of a nationwide malaria programme (Black 1956: 14). Anecdotal evidence from the Presbyterian Mission on the island of Tongoa in New Hebrides reported a resurgence of malaria following the abandonment of the war-time malaria control programme (Laing 1989: 146). Colonial officials made it clear that they would only use anti-malarial drugs sparingly after the war. A memorandum issued in 1944 stated 'When atebrin is given to natives by a white person there is a measure of control' and for this reason only small supplies of the drug could be 'safely entrusted to picked men – e.g. dressers or teachers'. The plan for the 'treatment of malaria in natives' in the post-war era was 'aimed at curing the attack rather than eliminating the parasite from the body' (in New Hebrides British Service 1944b). The endemic nature of malaria was once again emphasized as was the association of authority with 'whites'. It is evident that many New Hebrideans experienced both the sparse medical resources of the Condominium Government and the excesses of American war-time efforts.

Reducing malaria and finding a replacement for quinine became strategies of warfare. Mosquitoes were indiscriminate in their search for blood, attacking both Americans and the Japanese in the Pacific war (Patterson 2009: 154). Malaria, as Joy explained, could act as a 'force destroyer' when American and Allied forces were sick or as a 'force multiplier' when they were malaria free and the Japanese were not. The 'force multiplier' of malaria-free troops contributed to the defeat of the Japanese who suffered high rates of infection throughout the war in Melanesia (Joy 1999: 206). Edmund Russell argues that in the war against malaria, official and unofficial propaganda reinforced the tendency to view America's human enemies as pests and/or insects (2001: 114). Communication and educational materials crafted by the military aimed to inspire discipline in various ways, for example, by depicting the attacking mosquito as Japanese. The vector of malaria, the *Anopheles* mosquito, became militarized and racialized in new ways as war drove the science and discipline needed to control malaria (Patterson 2009: 154).

Conclusion

In 1925 as Buxton and Lambert travelled together throughout the archipelago on a reconnaissance mission to gather data about diseases such as malaria, particular scientific approaches to understanding biological and cultural differences as well as disease control travelled with them. Buxton was the first medical entomologist to point to the ways in which malaria eroded the health of Islanders and impeded demographic recovery from depopulation. Buxton undertook pioneering scientific research on the *Anopheles* mosquito, its habitat and its transmission of not only malaria but also filariasis, both of which were endemic in New Hebrides. However, as I have argued, Buxton, relied on 'naturalized' and 'pathologized' accounts of the inhabitants of New Hebrides and in so doing reproduced tropes of Melanesian primitivism. In this chapter, I have explored how racial thinking was reproduced in the science of disease control in colonial and military research. I tracked the connections between the specific locality of New Hebrides and the particular disease of malaria over two decades. The Second World War brought science and malaria together in new ways characterized by the singular focus on the *Anopheles* mosquito and the eradication of malaria in New Hebrides. While Buxton's work did not have direct effects on the bodies of New Hebrideans, as Vaughn has argued, the research and control measures enacted by the American entomologists and malarialogists did have effects, although unknown, on the bodies of New Hebrideans when they were medicated for malaria in the absence of a post-war malarial programme.

The adverse effects of malaria on the colonial and military populations and the need to secure labour 'accustomed to the tropics' (Rivers 1922: 89) were concerns that informed the British and French bids to create a plantation economy as well as the American war effort in New Hebrides in the 1940s. While 'racial thinking' was registered in the categories of civilized and primitive, 'native' and 'imported' races, reservoirs and carriers of disease, the category of labourer was overarching, construing New Hebrideans in particular ways as each new demand on their labour was made. As population steadily declined in New Hebrides, labour conditions became implicated in the failure of demographic recovery. The approach to labour changed again with the arrival of the Americans who needed indigenous labourers to help build their war apparatus.

Buxton's attention was drawn to the scientific problem of depopulation that signalled the failure of New Hebrideans to thrive. In contrast, the Americans underlined the Islanders' robust resistance to malaria through their acquired immunity that rendered them 'reservoirs of malaria'. American entomologists underlined the adaptability, efficiency and the capacity of the *Anopheles* mosquitoes to thrive in temporary and permanent breeding sites. This narrow focus on the mosquito effectively incorporated and reduced the bodies, the houses and subsistence practices of New Hebrideans to isolated elements and sites of research on the *Anopheles* mosquito. Research during the war provided an astonishingly rich vocabulary and choreography of the 'blooded' female *Anopheles* mosquito. The meticulous entomological research undertaken during the war, however, demonstrated that malaria was not just a biological phenomenon but rather was both natural and man-made. Engineers and entomologists generated man-made malaria during the war through their efforts to build bases and to destroy the mosquito habitat. Man-made malaria was also created in the work of plantations and by the Tonkinese labourers, 'the imported race' who grew water cabbage to supplement their meagre diets. Even more importantly, American entomologists pointed to the danger of the man-made malaria created by New Hebrideans as they conducted their day-to-day lives of growing taro, living near waterways and drinking water from wells. The scientific research of Buxton in the 1920s and that of the American entomologists in the 1940s, while vastly different in terms of scope and resources, relied on static notions and racialized explanations of differences that undermined the agency of Islanders and the Tonkinese The American entomologists correctly attributed considerable agency to the *Anopheles* mosquitoes but this attribution of agency appeared to eclipse that of New Hebrideans. In contrast, anthropological researchers such as Rivers provided a more complex and nuanced reading of New Hebrideans that privileged their agency while drawing attention to the adverse effects of colonial policy and practices as well as those of the plantation economy.

War-time deployments of science and discipline to combat malaria's vector, the *Anopheles* mosquito, and the plasmodium parasite, shaped the ways in which difference was construed through disease and race in New Hebrides. The intensive American research produced extraordinary data on the Anopheles mosquito whose deadly effects were transformed into a 'force multiplier' for the American and Allied troops and a 'force destroyer' for the Japanese. Difference in war-time New Hebrides was located in the complex assemblage of endemic and epidemic malaria, entomological and biomedical research, war-created environments, colonial legacies, American marines, indentured labour, indigenous 'reservoirs' and Japanese enemies. Differences among populations and environments came to be considered as scientific variables in the search for reliable labourers, healthy soldiers, and malaria-free environments.

Jean Mitchell is an Associate Professor of Anthropology at the University of Prince Edward Island, Canada. She has conducted research on gender, space, youth and health in several Pacific island countries in both colonial and postcolonial contexts.

Notes

1. Buxton's research on mosquitoes in New Hebrides was preceded by a brief research trip undertaken by M.A. Laveran who visited New Hebrides and published his research on mosquitoes in Efate, New Hebrides in 1902 (Laveran 1902).
2. Cooter and Sturdy argue that in public health programmes such as the Rockefeller Foundation, 'medicine served as a way of imposing the rationalizations of capital and philanthropy on native populations' (1998: 9).
3. While New Hebrides was not a theatre of active warfare, it was one of the principal forward bases for the American troops in the Guadalcanal Campaign in the nearby British colony Solomon Islands. The advance bases later became rear line support facilities (Laing 1989: 145; Lindstrom 1998: 397).
4. In Buxton's 1927 report he noted his research findings on filariasis (also known as elephantiasis). He found 31 per cent of 318 males over 12 years old from 16 islands were found to be *mf* positive indicating that microfilaria or the prelarval stage of the filarial worm was found. Furthermore, 6 per cent had elephantiasis, 7 per cent had enlarged scrotums, and 17 per cent had enlarged glands in the arm, see Buxton (1927: 1). The prevalence of filariasis varied from 0-11 per cent in different islands in the 1920s but had decreased by 1951, see Iyengar (1954). Also cited in Chevalier (n.d.: 1).
5. A 1942 article on malaria in the *American Journal of Nursing* also argued that 'In endemic regions many of the inhabitants, especially children, frequently show plasmodia in their blood although definite symptoms of malaria may be absent. These "carriers" are of great importance in the transmission of the disease' (Rausch 1942: 129).
6. Military historians have estimated that in the southwest Pacific from October 1942 to April 1943 there were ten allied soldiers admitted to hospital with malaria for every battle casualty (Beadle and Hoffman 1993: 322; Bennett 2006: 31; Webb Jr 2009: 157).
7. The bite of an infected female Anopheles mosquito introduces sporozoites into the bloodstream of humans and they then enter the host's liver. They undergo a nuclear and cellular division into numerous merozoites. A single sporozoite may generate from 10,000 to 30,000 merozoites. Five or six days later the merozoites enter the bloodstream and colonize the circulating red blood cells. It is at this stage that the parasite grows, consuming haemoglobin from the host cell. These new merozoites invade other red blood cells and this process of asexual reproduction is repeated every two or three days, depending on the species of malaria. The periodic blood cell invasion and bursting of the blood cells creates the human malaria symptoms of recurrent fevers and chills. See Packard (2007: 21).

Bibliography

Adams, R. 1990. 'Plantation Labour in Vanuatu', in C. Moore, J. Lecky and D. Munro (eds), *Labour in the South Pacific*. Townsville: James Cook University of North Queensland, pp. 140–163.

Anderson, W. 2006. *Colonial Pathologies: American Tropical Medicine, Race and Hygiene in the Philippines*. Durham: Duke University Press.

——— 2008. *The Collectors of Lost Souls: Turning Kuru Scientists into Whitemen*. Baltimore: Johns Hopkins University Press.

——— 2013. 'The Military Spur to Australian Medical Research', *Health and History* 15(1): 80–103.

Arnold, D. 1999. '"An Ancient Race Outworn": Malaria and Race in Colonial India, 1860–1930', in W. Ernst and B. Harris (eds), *Race, Science and Medicine, 1700–1960*. London: Routledge, pp. 123–143.

Baker, J.R. 1928. 'Depopulation in Espiritu Santo, New Hebrides', *Journal of the Royal Anthropological Institute of Great Britain and Ireland* 58: 279–303.

——— 1929a. *Man and Animals in the New Hebrides*. London: Routledge.

——— 1929b. 'The Northern New Hebrides', *The Geographical Journal* 73(4): 305–325.

——— 1935. 'Espiritu Santo, New Hebrides', *The Geographical Journal* 85(3): 209–229.

Beadle, C. and S.L. Hoffman. 1993. 'History of Malaria in the United States Naval Forces at War: World War One to the Vietnam Conflict', *Clinical Infectious Diseases* 16(2): 320–329.

Belkin, J., K.L. Knight and L.E. Rozeboom. 1945. 'Anophelene Mosquitoes of the Solomon Islands and New Hebrides', *Journal of Parasitology* 31(4): 241–265.

Bennett, J.A. 2006. 'Malaria, Medicine, and Melanesians: Contested Hybrid Spaces in World War II', *Health and History* 8(1): 27–55.

——— 2014. 'A Vanishing People or a Vanishing Discourse? W.H.R. Rivers' "Psychological factor" and Depopulation in the Solomon Islands and the New Hebrides', in E. Hviding and C. Berg (eds), *The Ethnographic Experiment: A.M. Hocart and W.H.R. Rivers in Island Melanesia*. New York and Oxford: Berghahn, pp. 214–251.

Black, R.H. 1956. 'The Epidemiology of Malaria in the Southwestern Pacific: Changes Associated with Increasing European Contact', *Oceania* 27(2): 136–142.

Buxton, P.A. 1926. 'The Depopulation of the New Hebrides and Other Parts of Melanesia', *Transactions of the Royal Society of Tropical Medicine and Hygiene* 19(8): 420–458.

——— 1927. 'Researches in Polynesia and Melanesia: An Account of Investigations in Samoa, Tonga, the Ellice Group, and the New Hebrides, in 1924, 1925', *Memoirs of the London School of Hygiene and Tropical Medicine* 1: 1–260.

Chevalier, C. n.d. *Sik Bigleg: Social Research in Filariasis*. Unpublished Report for the Pacific Regional Vector-Borne Disease Unit. Port Vila, Vanuatu: Secretariat of the Pacific Community, pp. 1–37.

Cirillo, V.J. 2011. '"Wonders Unconceived": Reflections on the Birth of Medical Entomology', *Perspectives in Biology and Medicine* 54(3): 381–398.

Condon-Rall, M.E. 1991. 'Allied Cooperation in Malaria Prevention and Control: The World War II Southwest Pacific Experience', *The Journal of the History of Medicine* 46(4): 493–513.

Cooter, R. and S. Sturdy. 1998. 'Of War, Medicine and Modernity: An Introduction', in R. Cooter, M. Harrison and S. Sturdy (eds), *War, Medicine and Modernity*. Thrupp and Stroud: Sutton Publishing, pp. 1–25.

Daggy, R.H. 1945. 'The Biology and Seasonal Cycle of Anopheles farauti on Espiritu Santo, New Hebrides', *Annals of Entomological Society of America* 38(1): 1–13.

Denoon, D. 1999. 'An Untimely Divorce: Western Medicine and Anthropology in Melanesia', *History and Anthropology* 11(2/3): 329–350.

Douglas, B. 2008. 'Foreign Bodies in Oceania', in B. Douglas and C. Ballard (eds), *Foreign Bodies. Oceania and the Science of Race 1750 – 1940.* Canberra: Australia National University E-Press, pp. 3–30.

Dureau, C. and M. Low. 1999. 'The Politics of Knowledge: Science, Race and Evolution in Asia and the Pacific', *History and Anthropology* 11(2-3): 131–156.

Fortun, K. 1993. 'Scientists and the Legacy of World War II. The Case of Operations Research (ORS)', *Social Studies of Science* 23(4): 595–642.

Gerber, E.J. 2008. 'Entomologists in World War II', in Armed Forces Pest Management Board Washington DC, *Proceedings of the DOD Symposium, 'Evolution of Military Medical Entomology',* Retrieved 12 July 2014 from http://www.dtic.mil/get-tr-doc/pdf?AD=ADA506261

Harrison, M. 2005. 'Science and the British Empire', *Isis* 61(1): 56–63.

Harrisson, T.H. 1936. 'The New Hebrides People and Culture', *Geographical Journal* 88(4): 332–341.

—— 1951. *The New Hebrides People and Cultures in New Hebrides Papers. Scientific Results of the Oxford University Expedition to the New Hebrides 1933-34.* Oxford University Exploration Club. London: Geoffrey Cumberlege, Oxford University Press.

Iyengar, M. 1954. 'Summary Data on Filariasis in the South Pacific', *South Pacific Commission Technical Paper* 132: 1–52.

—— 1955. 'Distribution of Mosquitoes in the South Pacific Region', *South Pacific Commission Technical Paper* 86: 1-47.

Joy, R.J.T. 1999. 'Malaria in American Troops in the South and Southwest Pacific in World War II', *Medical History* 43(2): 192–207.

Laing, J.K. 1989. *The Development of Medical and Health Services in New Hebrides and Vanuatu.* Auckland: Auckland Medical Historical Society.

Laveran, M.A. 1902. 'Sur Les Culicides des Nouvelles-Hébrides', *C.R. Société de Biologie Paris* 54: 908–910.

Lindstrom, L. 1998. 'Working Encounters: Oral Histories of World War II Labour Corps from Tanna', in G. White and L. Lindstrom (eds), *The Pacific Theatre: Island Representations of World War II.* Honolulu: University of Hawaii Press, pp. 395–418.

Lindstrom, L. and J. Gwero (eds). 1998. *Big Wok: Storian blong Wol Wo Tu long Vanuatu.* Suva: Institute of Pacific Studies, University of the South Pacific.

Link, M.M. and H.A. Coleman. 1955. *Medical Support of the Army and the Air Forces in World War II.* Washington: Department of the Air Force. Office of the Surgeon General USAF.

MacClancy, J. 1981. *To Kill a Bird with Two Stones.* Port Vila: Vanuatu Cultural Center.

MacLeod, R.M. 1999. 'Introduction: Science, Technology and the War in the Pacific', in R.M. MacLeod (ed.), *Science and the Pacific War: Science and Survival in the Pacific 1939-1945.* Dordrecht: Kluwer Academic Publishers, pp. 1–12.

Meyerhoff, M. 2002. 'A Vanishing Act: Tonkinese Migrant Labour in Vanuatu in the Early Twentieth Century', *The Journal of Pacific History* 37(1): 45–56.

Mitman, G., M. Murphy and C. Sellers. 2004. 'Introduction: A Cloud Over History', *Osiris – Special issue Landscapes of Exposure: Knowledge and Illness in Modern Environment* 19: 1–17.

New Hebrides British Service. 1944a. 'Pamphlet on Malaria and Malaria Control', Commander, Espiritu Santo Island Command, Navy 140, WPA NHBS 8/11/ F8/2/1d. 1–3. (Unpublished Material from the Western Pacific Archives [WPA], University of Auckland, New Zealand.)

New Hebrides British Service. 1944b. 'Notice. Use of Atebrin. Memorandum', Office of the British Resident Commissioner, WPA, NHBS 7/XI/7. 1–2. (Unpublished Material from the Western Pacific Archives [WPA], University of Auckland, New Zealand.)

Opeskin, B. 2009. 'Malaria in Pacific Populations: Seen But Not Heard', *Journal of Population Research* 26(2): 175–199.

Packard, R.M. 2007. *The Making of a Tropical Disease: A Short History of Malaria.* Baltimore: The Johns Hopkins University Press.

Patterson, G. 2009. *The Mosquito Crusades: A History of the American Anti-Mosquito Movement from the Reed Commission to the First Earth Day.* New Jersey: Rutgers University Press.

Perry, W.J. 1945. *Notes on the Biology of the Malaria Vector in the New Hebrides – Solomon Islands. An Information Guide for all Entomologists.* Headquarters Malaria and Epidemic Disease Control, South Pacific Area. Retrieved 12 July 2014 from http://www.mosquitocatalog.org/files/pdfs/100299-1.pdf

_____ 1946. 'Keys to the Larval and Adult Mosquitoes of Espiritu Santo (New Hebrides) with Notes on Their Bionomics', *The Pan-Pacific Entomologist* 22(1): 9–19.

Rausch, L.E. 1942. 'Malaria', *The American Journal of Nursing* 42(2): 124–132.

Rivers, W.H.R. 1922. 'The Psychological Factor', in W.H.R. Rivers (ed.), *Essays on the Depopulation of Melanesia.* Cambridge: Cambridge University Press, pp. 84–113.

Ross, R. 1923. *Memoirs with a Full Account of the Malaria Problem and its Solution.* London: John Murray.

Russell, E. 2001. *War and Nature: Fighting Humans and Insects with Chemicals from World War 1 to Silent Spring.* New York: Cambridge University Press.

Shineberg, D. 1999. *The People Trade: Pacific Island Labourers and New Caledonia 1865-1930.* Honolulu: University of Hawaii Press.

Shlomowitz, R. 1992. 'Differential Mortality of Asians and Pacific Islanders in the Pacific Labour Trade', *Journal of Australian Population Association* 7(2): 116–127.

Slater, L. 2004. 'Malaria Chemotherapy and the "Kaleidoscopic" Organization of Biomedical Research during World War II', *Ambix* 51(2): 107–134.

_____ 2009. *War and Disease: Biomedical Research on Malaria in the Twentieth Century.* New Brunswick: Rutgers University Press.

Speiser, F. 1922. 'Decadence and Preservation in the New Hebrides', in W.H.R. Rivers (ed.), *Essays on the Depopulation of Melanesia.* Cambridge: Cambridge University Press, pp. 25–61.

Spencer, M. 1992. 'The History of Malaria Control in the Southwest Pacific Region, with Particular Reference to Papua New Guinea and the Solomon Islands', *Papua and New Guinea Medical Journal* 35(1): 33–66.

Stapleton, D.H. 2005. 'A Lost Chapter in the Early History of DDT: The Development of Anti-Typhus Technologies by the Rockefeller Foundation's Louse Laboratory, 1942–1944', *Technology and Culture* 46(3): 513–540.

Thomas, N. 1990. 'Sanitation and Seeing: The Creation of State Power in Early Colonial Fiji', *Comparative Studies in Society and History* 32(1): 149–170.

_____ 1994. *Colonialism's Culture: Anthropology, Travel and Government.* Princeton: Princeton University Press.

Thompson, R.C. 1981. 'New Hebrides Frontier Natives and Settlers on the New Hebrides Frontier 1890-1900', *Pacific Studies* 5(1): 1–18.

Tilley, H. 2004. 'Ecologies of Complexity: Tropical Environments, African Trypanosomiasis and the Science of Disease Control in British Colonial Africa 1900-1940', *Osiris – Special Issue – Landscapes of Exposure: Knowledge and Illness in Modern Environments* 19: 21–38.

Vaughn, M. 1991. *Curing Their Ills: Colonial Power and African Illness*. Redwood City: Stanford University Press.

Walker, A.S. 1952. *Clinical Problems of War*. Canberra: Australian War Memorial.

Webb Jr., J.L.A. 2009. *Humanity's Burden: A Global History of Malaria*. Cambridge: Cambridge University Press.

White, L. 1995. 'Tsetse Visions: Narratives of Blood in Bugs in Colonial Northern Rhodesia 1931-1939', *The Journal of African History* 36(2): 219–245.

Widmer, A. 2008. 'The Effects of Elusive Knowledge: Census, Health Laws and Inconsistently Modern Subjects in Early Colonial Vanuatu', *Journal of Legal Anthropology* 1(1): 92–116.

——— 2012. 'Of Field Encounters and Metropolitan Debates: Research and the Making and Meaning of the Melanesian "Race" during Demographic Decline', *Paideuma* 58: 69–93.

——— 2013. 'Seeing Health Like a Colonial State: Assistant Medical Practitioners and Nascent Biomedical Citizenship in the New Hebrides', in S. Trnka, J. Park and C. Dureau (eds), *Senses and Citizenships: Embodying Political Life*. New York: Routledge, pp. 200–220.

——— 2014. 'The Imbalanced Sex Ratio and the High Bride Price: Watermarks of Race in Demography, Census, and the Colonial Regulation of Reproduction', *Science, Technology, & Human Values* 39(4): 538–560.

4

Medical Missions – Racial Visions

Fighting Sleeping Sickness in Colonial Africa in the Early Twentieth Century

Sarah Ehlers

> Knowledge has accumulated. Fighting the sleeping sickness is like [s]laying a vampire. To make the spell work, five separate conditions must be present – water, bushes, trees, the tsetse-fly (*Glossina palpalis*), and one infected person. Remove any of these conditions and the curse is lifted. But let them all be conjoined, and the sure destruction of every human being in the district is only a matter of time.
>
> —Winston Churchill, *My African Journey*

At the beginning of the twentieth century, Winston Churchill's *My African Journey* received much attention and sparked the imagination of its readers at home. But as well as providing narratives of adventures and exoticism, *My African Journey* also offers a specific imperialist take on the British Protectorate Uganda as well as a broader blueprint for the future of Empire. The application of scientific knowledge formed an integral part of Winston Churchill's and others' colonial visions for the future.[1] However, that knowledge did not accumulate by itself but was – in the case of sleeping sickness – created by European doctors who found themselves at the heart of the colonial project. The aim of this chapter is to explore in which ways the pursuit of this knowledge connected with the colonial context, more specifically, with governmental penetration and explicit or veiled concepts of race.

Churchill's interest in the medical situation in Uganda was driven by a genuine commitment. Being close to the British actors of the sleeping sickness campaigns, he accompanied doctors as they screened the African population or discussed the matter with scientists and administrators (Churchill 1989: 50–72). Rife with positive visions for the future, his travel depictions reveal the hope for implanting civilized virtues among the colonial subjects by developing Uganda through the means of modern technology and indirect rule. In this regard, understanding tropical diseases and erasing epidemics was as crucial for the desired transformation as for the colonial self-image itself. Indeed, Churchill's literary accounts of the fight against sleeping sickness not only construct a rational interplay between scientific, administrative and medical measures but also heighten the otherness of Africa, its nature and its people, in comparison to Europe and the perceived civilized colonial power. The quoted passage thereby represents a larger literary tradition that is basically fuelled by the clash of two spheres: it depicts a world of danger, witchcraft and hostile nature that is penetrated by Western science, rising to fight superstition, the maladies and misery of the past.[2]

In her recent analysis of popular medical writing on colonial Africa, Anna Crozier identified certain rhetorical tropes 'that juxtaposed western ingenuity against racialized constructions of Africa and the African people as inherently pathological' (2007: 393). My chapter deals with medical writing that was not directed at a broader audience but served as internal communication between doctors, researchers and bureaucrats. Following the knowledge that travelled between European scientists and colonial authorities, I work with what created a European frame of reference for the practitioners of colonial sleeping sickness campaigns: distinguishing between Africans and Europeans, not between local populations, their reports assemble and compare cases from all over Africa. By exploring how notions of racial difference have been implicated in research constellations as well as in administrative attempts to fight the disease, I aim to reveal the racialized dimension of colonial sleeping sickness campaigns. First, my chapter discusses how the conceptions of sleeping sickness varied depending on the patient's race. How did colonial doctors perceive their African patients, how did they react to infected Europeans? How did their depiction of the symptomology differ, how did the perceived patient's race impact on the medical treatment? Second, I focus on prevention and containment measures enacted during the sleeping sickness campaigns. Distinguishing between African and European inhabitants, the campaigns offer a rich source for exploring different takes on populations at risk. Third, I discuss representations of race as a component of environmental interventions during the sleeping sickness campaigns. In conclusion, this chapter sheds light on the ways in which racial thinking was structured and reaffirmed by daily medical practice in sleeping sickness areas.

It comes as no great surprise that race surfaced as an important component of dealing with an infectious disease in a colonial context, not only in the quoted image of the sleeping sickness spell but also in the actual medical campaigns against it. The 'one infected person' in Churchill's model represents an African who was considered as one variable among other natural elements. In reality, this single person was one of hundreds of thousands of infected Africans surrounded by masses of water, bushes, trees and tsetse flies, as elaborated on below.

Sleeping Sickness – a Distant Disease

Sleeping sickness (*Human African Trypanosomiasis*) is an infectious disease only prevalent in tropical Africa. It is caused by a protozoan parasite (*Trypanosoma gambiense* or *rhodiense*) and transmitted by the bite of the tsetse fly. If left untreated, it is eventually fatal (WHO 2006). While its germ and its carrier were identified within the first years of the epidemic, a medication was not developed until the 1920s.[4]

Although the disease had been known in Africa for a long time, its epidemics at the beginning of the twentieth century sent shock waves around the colonial metropolises in Europe. The degree to which these epidemics were caused by colonial penetration is still a matter of controversy. However, it is beyond dispute that warfare, and ecological and social disruption fostered its spread. As a result of migration, flight, and also advanced traffic networks, people abandoned their homes, crossed uninhabited areas and settled in new territory. Thus, the infection spread to vast regions whose inhabitants had been previously unaware of the disease.[5] With the death toll reaching hundreds of thousands, and the progression of the disease going unchecked, sleeping sickness created a situation on colonial territory that the administrations were unable to ignore. In the years to come, all concerned European colonial powers developed campaigns against the disease, combining immense medical efforts with authoritarian and sometimes brutal approaches towards the affected indigenous population. The measures introduced in fly-infested zones ranged from evacuating populated areas, clearing the habitat of the tsetse fly from bushes, trees and swamps, closing roads and landings, performing mass screenings of Africans, establishing medical checkpoints for the African population and sleeping sickness camps for the infected, and testing possible medications.[6]

However, whilst epidemiological or governmental intentions were distinguishable, the line between their outcomes was severely blurred. Pointing in this direction, Maryinez Lyons coined the term 'the colonial disease' to describe the interwovenness of the fight against the disease and colonial politics. Indeed, as Lyons (1992) clearly demonstrates, affected Africans perceived the medical campaigns as aggressive expressions of colonial rule. Besides their immense impact on African societies, sleeping sickness campaigns also altered the affected landscape: destroying the tsetse habitat meant an ongoing restructuring of the African environment.

At the beginning of the twentieth century, tropical medicine was not the only domain steeped in ideas of helpless Africans and superior European science.[7] But even compared to other areas of colonial medicine, sleeping sickness research took an exceptional position. With the symptomatology of sleeping sickness evoking ideas of passivity and lethargy, depictions of a distinctly African disease pattern flourished in the scientific journals, and beyond. This was particularly true in the first years of the epidemic when sleeping sickness was conceived of as an infection affecting only Africans. Even when it became known that Europeans were also infected, thereby debunking this theory, distinct clinical pictures of Africans prevailed and led to distinct medical measures. In assessing German, French and British reports, it is striking how similar they were. Although sleeping sickness affected a vast territory and the doctors' reports and the doctors themselves came from different places, they reveal a common ensemble of representations that was structured along the categories 'European' and 'African'.

Indifferent Victims

Basically, two motifs on Africans run through the reports. Although African patients were depicted as apathetic and helpless, doctors warned continuously of the danger posed by their mobility and unruliness. At the same time that they provided accounts of mass death, doctors pointed out that Africans were unable to counter this situation by themselves. 'When once trypanosoma is introduced [the natives] die like rotten sheep', the *New York Times* quoted David Bruce as saying in 1908, linking human trypanosomiasis with epizootic diseases.[8] Emphasizing passiveness, however, did not imply denying African responsibility in spreading the disease. In particular, Europeans dealing with administrative containment measures complained about the indifference of Africans in the face of the epidemic. 'This vast territory is occupied by natives who are apathetic and will not voluntarily lift a finger to save themselves from annihilation' (Sleeping Sickness Bureau 1909c: 106), the London Sleeping Sickness Bureau concluded in 1909. Indifference did not only mean refusing anti-sleeping sickness measures but also referred to physical components. Colonial doctors often remarked that Africans overlooked the bites of tsetse flies or reacted with ignorance.[9] In their accounts, it was specifically Africans who were frequently bitten by tsetse flies, due to their being overexposed to them – 'given their lack of clothes' (Meixner 1910: 261). Some doctors pointed out that people living in fly-infested areas would get used to fly bites and therefore would not take much notice of them. Thus, the African was required to learn to take responsibility for his body and become accustomed to 'European needs', for instance, wearing clothes.[10] Complaining about African indifference was also part of a repeated narrative about the alleged lack of hygienic consciousness among the colonized (Sleeping Sickness Bureau 1909c: 115). German scientist Hans Ziemann, for instance, pointed out that sleeping sickness was most powerful in 'degenerated tribes', concluding: 'therefore we must do anything in order to improve the population socially and hygienically' (1912: 139).[11] The French researchers Martin and Lebœuf also asserted a general lack of health awareness among Africans: the patients they examined stated that they felt not the slightest malaise while their body temperature was beyond 39°C (1908: 387).[12]

Investigations of the medical missions usually followed the same pattern: doctors visited the infected areas, reported the scale of the epidemic to the colonial authorities and detained the infected Africans. The colonial doctors' attention was not only focused on infection rates but also on social structures of African society: the names and size of villages, their economical basis, age, gender and livelihood of their inhabitants. They reported their findings to the colonial authorities.[13] Inside the sleeping sickness camps, the patients could be observed at close range and different forms of treatment could be explored. To this end, patients had to wear a badge around their neck displaying their identification number, medication and application rate.[14] Serving both research and curative aims, doctors were interested in obtaining comparable results. Although doctors continuously struggled to keep patients under treatment, sleeping sickness camps enabled them to observe the effects of a given medication on a great number of cases and for a significant amount of time. When it came to publishing the results, their journal articles included numerous tables with the patients' numbers, age, sex, weight, medication, doses, treatment results, beginning and end of treatment which could refer to the patient's improvement, flight from the camp or the patient's death.[15]

Endangered Europeans

It was as late as December 1903 that colonial doctors became aware that sleeping sickness also threatened Europeans. This change of perception was due to the death of so-called Mrs S., a British woman who accompanied her husband, a missionary, to the French Congo. Her name, Mrs S., was created when she became famous in the medical world for being the first European to succumb to sleeping sickness. Up to her death, sleeping sickness was perceived as a genuinely African disease, not only confined to the African continent but also to black people. Although the disease's infectious character was an established fact, in medical dictionaries it still surfaced as 'negro lethargy', 'African sleeping sickness', 'maladie du sommeil des nègres', 'narcotisme des nègres' or 'Schlafkrankheit der Neger'.[16] While the nature of the perceived 'racial immunity' was discussed only rarely in scientific journals, many doctors remarked that obviously sleeping sickness targeted solely Africans.[17] When Robert Koch asked for research funding from the German health ministry to investigate sleeping sickness in the German colonies, understanding this somewhat contradictious nature of infection was one of his scientific aims: 'It is peculiar that only the black population suffers from it [sleeping sickness] but is able to spread the disease' (1902).

Given this established perception, it is no surprise that it took several months to understand the nature of Mrs S.' malady. Diagnosed with malaria, she returned to England where she was examined by several luminaries of British tropical medicine. However, nobody identified her infection, let alone was able to cure her. The day after she died, Patrick Manson, head of the London School for Tropical Medicine, published an article in the *British Medical Journal.* Under the title 'Sleeping Sickness and Trypanosomiasis in a European: Death: Preliminary Note', he presented the first results of the post-mortem and, in doing this, the first proof for sleeping sickness in a European. Manson was so preoccupied about her death that he felt the need to inform the medical profession immediately: 'Considering that this is the first well-authenticated case of sleeping-sickness in a European [...] and that the subject is deservedly exciting much interest, I feel justified in offering this preliminary note for publication' (1903: 1461).

Mrs S.' diagnosis caused a stir within the medical profession and beyond. Given her exceptional status, her case was a fascinating topic and international medical journals reviewed Manson's publication and the follow-up articles (ASTH 1904).

Her brain was exhibited at the London School for Tropical Medicine, giving visitors the unique opportunity to study the effects of sleeping sickness on the European body (Fülleborn 1904: 289). However, after a few months' time when colonial doctors were aware of the possibility of infection, her case was no longer unique (Manson 1908: 35; Dutton, Todd and Christy 1905: 93).

When infections of Europeans became public, this had a twofold impact. With regard to health measures, the assessment of risks for Europeans in affected areas changed significantly. Although sleeping sickness was still framed and imagined as an African disease, in the mindsets of the colonizers it was transformed from an epidemic threatening the colonial subjects – and along with that the colonial economy – to a danger also targeting individual colonialists. In the realm of research, a new type of medical reports arose, now dealing with sleeping sickness in Europeans. Those widely published descriptions of European patients were not absorbed into the bulk of existing research on African patients, but rather presented as individual case studies with special attention given to gender and class issues.[18]

Although Europeans infected with sleeping sickness were rare, they were the object of considerable attention and formed a distinctive genre in the scientific journals. In discussing every European infection as an individual case, giving information on age, gender, profession and details on the situation of infection, the reports on Europeans differ clearly from those on Africans.[19] An interest in infected Europeans united all European research institutions. While the reports gave no specific reasons why European and African infections were discussed separately, it becomes clear that in terms of treatment goals and side effects, reports on Europeans were used as references for other infected Europeans – although much more research existed on infected Africans (see, for example, Manson 1908: 33).

European patients were separated from Africans not only in the journals but also in practice. Wherever resources sufficed, two parallel structures emerged: while Europeans were generally treated at home or in a European hospital, Africans were sent to sleeping sickness camps.[20] However, segregated patients were often examined by the same doctors. For instance, Vaughan describes the schedule of Hugh Stannus who worked in Nyasa during the sleeping sickness epidemic. After having visited African patients in the native hospital, he continued to the European hospital and eventually did his home visits with Europeans (Vaughan 1991: 29–30).

After the first European infections became known, all colonial powers published protection guidelines. The London Sleeping Sickness Bureau, for instance, advised that native water carriers should not be allowed to enter European houses and 'Europeans should be on the lookout for the disease in their servants, for there can be no doubt that the constant presence of an infected person is a danger' (1909d: 9). A German guideline referred to the internationally proven advantage of white clothes for Europeans in affected areas and added: 'If he is in a fly area, the European has to be accompanied at least by one, if possible even two boys equipped with fly or butterfly nets and trained in catching flies. It is very practical if one of the boys wears a black coat, because it attracts flies which makes them easier to catch' (Anonymous 1909, translation: S.E.).

Regarding the safety of Europeans in the colonies, women took an exceptional position in the minds of the colonizers. In medical reports on sleeping sickness, several doctors emphasized the high liability of women to the disease: given the small number of females present in the colonies, colonial doctors were surprised by the high number of infected women. Their surprise was increased by the common observation that European women in the colonies were hardly found outside and were therefore a difficult target for tsetse flies (Manson 1908: 50; 1910: 72; Sleeping Sickness Bureau 1909a: 96f; 1909b: 486; 1912: 18). In an article on European women and sleeping sickness, Patrick Manson reflected on the fact that the majority of women had their fly bites on their – supposedly well protected – legs: 'One would suppose that the petticoat would afford a protection even more effective than the trouser does in man' (1908: 51). The London Sleeping Sickness Bureau also weighed in, elaborating on the perspective of the tsetse fly: '[I]t prefers dark to bright spots; it is therefore not unlikely to get under a petticoat' (1909b: 486).

Writing about endangered women in the tropics has often been interpreted as a form of othering: male colonial actors describing women as weak and threatened implicitly secured their own position as strong and safe. Indeed, these comments also fit into a broader concept that declared women as unfit to stand the hardships of tropical climate or – since the turn of the century – at least emphasized its danger for the female body.[21]

But beyond revealing the maleness of colonial medical discourse, these comments also mirror the racialized dimension of sleeping sickness research. Worrying about white women created a subcategory that further distinguished Europeans from Africans. Since doctors paid no attention to gender difference regarding African sleeping sickness patients, no articles were published on the susceptibility of African women to the disease. European women had no counterpoint in the collective of African patients.

The colonial application of European as an identity category is interesting for two reasons. First, the sleeping sickness campaigns were not only about protecting the nation's members. Instead, for instance when German Regierungsrat Professor Beck reported to the German Colonial Office that 'as many as ten Europeans at Tanganyika had come down with sleeping sickness' (1909), the concern about a racial collective becomes apparent. Also, most of the protection guidelines explicitly addressed Europeans and not national collectives. Second, the category Europeanness included biological as well as cultural, political and economic components, for instance in demanding that Europeans should be vigilant against possible infections of their servants and workers or referring to 'European needs'. In every instance, these components were bound to physical traits: Black 'boys' in black coats were supposed to hold tsetse flies off Europeans dressed in white.

The assessment of risks for Europeans in sleeping sickness areas changed significantly in the first years of the epidemic. This radical shift from an image of the immune to the endangered European can be illustrated with the following example. In the first years of the epidemic, boats were equipped with wired cages to protect local workers from fly bites for their safety and to prevent possible infected Africans from being bitten, while the boat was a space of free movement for Europeans. This practice was common in French, British and Belgian infected areas, but changed when European infections emerged.[22] In 1909, the London Sleeping Sickness Bureau published a guideline that said: 'Steamers which ply on fly-infected rivers should be provided with wire-gauze cages into which Europeans can retire' (1909d: 8).

Disseminators of Infection

The European perception of Africans as dangerous was particularly linked to African mobility. First, this was grounded in identifying Africans as disease spreading elements. Second, the prevailing idea of Africans as uncontrollable was also reflected in the reports. Confining the epidemic by isolating the infected was one of the first attempts of all colonial powers to enact an anti-sleeping sickness policy. This measure targeted only Africans while European freedom of movement was never touched – even if an infection was detected. Preventing the mobility of infected Africans, however, generated severe difficulties: how to distinguish between the infected and non-infected?

In 1903, the Uganda administration began to prevent all Africans showing sleeping sickness symptoms from using the railway system. One year later, Robert Koch reported on the British experiences, where due to the long incubation period of sleeping sickness, the ban had been proved ineffective. Excluding only those passengers who already reached the final stage was not enough to stop the epidemic from spreading (1904: 194). In this complaint, Koch highlighted a feature that gave sleeping sickness its mysterious and dangerous character: sleeping sickness infections were externally invisible in the first stages. Scientific journals frequently presented cases of Africans whose infections remained externally invisible for years, thereby posing a problem not only for the colonial authorities.[23] For instance, Jeanselme and Rist, in their manual of exotic pathology, highlighted the fact that Africans were particularly known for neither showing any symptoms nor complaining about any, even during long-lasting infections (1909: 91). Thus, infections and stages of the disease could only be determined by microscopic analysis that was possible since the sleeping sickness germ had been identified in 1902, but required equipment and medical knowledge.

In writing about apparently healthy Africans whose dangerousness revealed itself only under the microscope, doctors rearticulated and reasserted common stereotypes about hidden dangers in African nature.[24] This tradition not only shaped travel writing and literature but also the perceptions of colonial doctors. When Robert Koch portrayed his African oarsmen – 'young, strong men who almost continuously oared for 12½ hours' – he added: 'By the way, out of these 52 seemingly healthy persons, 47 had filarial perstans, 26 malarial parasites and two relapsing fever spirilla in their blood. This is what the still considered healthy population looks like' (1912c: 545).[25] The reports on sleeping sickness do not contain any accounts on physically healthy but infected Europeans.

Confining hidden infections to Africans seems somehow plausible if the colonial doctors' routine is taken into account. Doctors involved in sleeping sickness campaigns spent most of the time travelling to infected areas and examining the African local population by analysing their blood and lymph fluids.[26] The number of Africans they examined was enormous. After having examined 25,865 Africans in North Nyasa District in 1910, Dr Davey summarized his findings thus: 73 per cent of the adults and 94 per cent of the children were infected with sleeping sickness (Hearsay 1910: 167). Mass examinations like these confirmed Robert Koch's portrayal of his oarsmen: the majority of healthy looking Africans turned out to be infectious trypanosome carriers. Looking for infected Africans was part of the doctor's daily routine that involved the view that every African encountered became a possible sleeping sickness carrier. Grounded in such perceptions, the ensuing segregation policy did not distinguish between infected and non-infected but between Europeans and Africans.[27] Colonial town planning also implemented this paradigm: fighting sleeping sickness involved separating European from African neighbourhoods.[28] In addition, sleeping sickness camps provided for spatial separation between Europeans and Africans, including the non-infected African staff.[29]

Given the geographical distribution of sleeping sickness, it comes as no great surprise that Africans were primarily considered as trypanosome carriers. In comparison, the numbers of infected Europeans were negligibly small.[30] For all colonial powers, however, engaging with sleeping sickness had a decisive bearing on the dynamics of colonial identities and created the motif of Africans as disease carriers.

Restricting African Mobility

A testament to the potency of this perception can be found in travel restrictions enacted during the sleeping sickness campaigns.[31] As mentioned above, African mobility was considered as the chief cause for spreading the disease and was thus increasingly circumscribed by the legislation.[32] Referring to sleeping sickness, the Cameroon administration attempted to prevent Africans coming from French Equatorial Africa (FEA) from entering or at least to keep them under strict surveillance. Medical passports, attesting to the fact that the traveller was found uninfected and valid for fifteen days, were to be issued for crossing the borders.[33] In 1908, Dr Feldmann asked the Gouvernement Cameroon to persuade the Congo administration to pay more attention: after examining sixteen Africans with valid medical passports, he found ten of them infected (Feldmann: 1908b). The Belgian Congo introduced a *cordon-sanitaire* policy in 1906 while Uganda opened only a certain number of roads and landing places to traffic in infected areas (Hoppe 2003: 55). In FEA, using waterways and main roads required medical certificates (Headrick 1994: 90).[34]

Both doctors and involved officials complained about the lack of African sedentary settlements.[35] In 1907, medical officer Kudicke reported on the problems of the sleeping sickness campaign in Kigarama (German East Africa, GEA), claiming that the disease could only be prevented from spreading if African mobility was restricted.

It was of secondary importance for Kudicke if those travellers were infected; his first concern was to limit every instance of African mobility. He added: 'As I understand it this will be difficult. [...] Thus, surveillance of travelers is required' (1907: 18-19).

Attempts to restrict African mobility were not without controversy. A colonial official from Ujiji (GEA) discussed the fact that travel restrictions to Rhodesia only applied for Africans but not for infected Europeans. In that case, he added, the restrictions were untenable (Governor of German East Africa 1908: 142). *The Journal for Tropical Medicine* questioned particularly their outcome by asking: 'To what purpose would be any international agreement as to the sanitary regulation of travelers in a country where, while there are no regular roads, the entire jungle is a maze of paths formed by hippopotamus, rhinoceros, and other large game, the practicability of which for human intercourse is known only to the natives?' (Anonymous. 1908. 'Professor Koch's View : 56). Here, African mobility was not only linked with the concern of spreading the epidemic, but also referred to the Europeans' perceived insecurity. The image of a jungle where only Africans were able to orient themselves contrasts sharply with the claim of possessing the scientific instruments necessary to colonize a continent.

Enforcing travel restrictions also brought with it the risk of failure. The likelihood of this risk reveals itself in the doctors' reports. German colonial physician Neubert, for instance, demanded in 1906 that travel restrictions be lifted in the area of Ujiji (GEA) because they obstructed the local economy. He argued that the restrictions only served to foster contraband trade and had no effect on avoiding the danger of spreading the disease.[36] Frequently doctors reported the difficulties of preventing African workers from visiting their families.[37] Apart from the fact that travel restrictions were often unenforceable and therefore undermined colonial authority, they posed a second problem. With local trade and traffic being dependent on mobile workers, travel restrictions caused conflicts, as Neubert's petition suggests, between doctors and the local economy. Critics, however, did not limit themselves to economic aspects: Lyons quotes the example of Dr Zerbini who objected to the Congo sleeping sickness policy. Even in cases of highly infectious tuberculosis, such a deprivation of liberty would be unthinkable in Europe, he argued (1992: 110).

Visions of Nature

Epidemics and the responses to them play a major role in African colonial history. Medical control schemes never targeted the affected population alone but created technocratic openings for the articulation of colonial visions of the African environment. Their implementation had a considerable impact on local ecosystems and nature itself as in the minds of the colonizers, extinguishing diseases was inextricably linked to environmental action.[38] Flies, and insects in general, received broad attention among researchers of tropical medicine at the beginning of the twentieth century. When first defining tropical diseases, Patrick Manson highlighted the importance of insects as alternate hosts. Meanwhile Ronald Ross, discoverer of the mosquito transmission of malaria, seeking ways of controlling the disease on a global scale, proposed the eradication of anopheles mosquitoes. The more tropical medicine was applied in the field, the more controlling the parasitic vectors became a leitmotif of disease control.[39] This was particularly true for sleeping sickness since – unlike malaria – there were no available cure till the late 1920s, leaving prevention as the only viable option.[40]

Although some researchers asked for the extinction of the tsetse fly, colonial administrations were generally aware that this would border on the impossible (Bruce 1908: 257). Instead, they favoured altering the environment, thus attempting the destruction of the flies' habitat. Especially in British and German infected areas, creating fly-free zones by bush clearing and tree thinning became a major goal of sleeping sickness policy that was mainly deployed in areas of strategic importance such as traffic junctions, landings and European neighbourhoods. In French West Africa, strategic clearing was also a common demand but due to a general lack of resources, colonial administrators failed to put their concepts into practice. The German and British colonial administrations shared an interest in animals as possible reservoirs for infection. While British research focused mainly on large game, German scientist Robert Koch developed the idea that crocodiles acted as reservoirs and should therefore be exterminated in fly areas.[41] On a large scale both schemes, attacking tsetse habitats and reservoirs of infection, turned out to be a failure. Clearing limited areas only provided limited safety and did not confine the disease. The same applied for decimating host animals. Robert Koch, for instance, complained that the crocodiles' cadavers only attracted new crocodiles, which made the situation even worse. Thus, he proposed to abandon 'this far too small scale' (1912a: 948) of environmental interventions.

Despite all these activities, sleeping sickness researchers were well aware that the principal reservoir of trypanosomes was man. Depopulating infected areas was a principle common health measure carried out around Lake Victoria and Tanganyika, but also in other areas. The idea behind sleeping sickness depopulations was to temporarily separate people from their infectious environment. In theory, this would eventually lead to trypanosome free fly populations as David Bruce explained to the Royal Society in 1908:

> It is evident that if we remove all the affected natives from the region of the fly, we at once stop the occurrence of fresh cases. Where there is no fly there is no sickness. And as the fly is only infective for 48 hours, it is also evident that you could repopulate with safety the sleeping sickness area – that is the 'fly region' – at the end of that time with a healthy population from the interior of Uganda. (1908: 257)

Such planning reflects a way of thinking which is out of touch with local conditions; it is technocratic and increasingly disconnected from the needs of the local population. Coining the term of 'authoritarian high modernism', James Scott analysed the role of mapmaking and state simplifications, emphasizing the transformative character of maps:

> [Maps] are designed to summarize precisely those aspects of a complex world that are of immediate interest to the mapmaker and to ignore the rest. [...] This transformative power resides not in the map [...] but rather in the power possessed by those who deploy the perspective of that particular map. (Scott 1998: 87)

In such a perspective, the African population was increasingly perceived as a – potentially infected – environment variable, which could be displaced according to the needs of health planners. Indeed, social engineering, environmental takes and disciplinary action in the context of public health measures formed an intertwined matrix, which constituted an integral part of high colonialism. However, ambitious planning and technocratic schemes were more often than not undermined by reality: like many other colonial projects, its protagonists had to cope with frustrations and failure. Bush clearing, tree thinning and game destruction did not provide security for larger areas, people removed from tsetse areas managed to come back and, in general, separating humans and tsetse flies from another tended to be nigh on impossible.[42] In evaluating the poor outcomes of their campaigns, colonial doctors only rarely pointed to the shortcomings of their schemes. Instead, they blamed the human element as the most unpredictable and unreliable. In this regard, complaining about Africans for crossing tsetse areas or not taking care of already cleared spots not only blurred the line between humans and nature but also held Africans responsible for the dangers of their natural environment. For instance, when explaining the ineffectiveness of environmental interventions such as bush clearing in German East Africa, doctors stated that their 'efforts were destroyed by external conditions or native stupidity' (Kleine 1915: 97),[43] hence demonstrating the fact that they saw the local population as part of the problem rather than as part of the solution – and certainly not as potential partners.

Conclusion

For scientists and medical officers as well as for colonial authorities, researching and controlling sleeping sickness provided the opportunity to articulate and enact visions of colonial Africa. Central to these visions was not only the perception of Africa as a hostile environment to be transformed, but also of its African population as a part of nature and – along with that – as an ensemble of disease spreading elements that could be subjected by deploying control measures. Knowledge of the colonial situation and its respective needs was a major component of medical thinking in areas at risk and gave rise to racialized concepts of disease. Defining Africans' relationship with their environment, medical thinking during this period was grounded in binarisms of health and sickness, hostile and endangered. With growing difficulties in controlling the disease, medical campaigns targeted not patients but Africans in general. Furthermore, when it came to restructuring the affected landscape, Africans were held responsible for the disorder of their environment that harboured the tsetse fly. Coping with the disease can therefore be interpreted as a typical example of social and environmental engineering.

Campaigns against sleeping sickness provide a telling illustration of how conceptions of racial difference permeated medical thinking. The segregation of African and European patients took place on many different levels. First, there was a spatial segregation since they were treated in different places. Whilst infected Europeans – if their condition allowed it – left Africa for their home country, doctors tried to gather African patients in sleeping sickness camps. Inside these, therapeutic approaches were large-scale oriented and did not leave much room for individual modification. Second, depending on the patient's race, the nature of infection was imagined in a different way as they were portrayed as neither displaying nor noticing any signs of infection. The pathologization of Africans went hand in glove with connecting sleeping sickness symptoms to their bodies, their way of life and their character. It is in these terms that the socialization of racialized medicine in colonial Africa may be observed. Although Europeans and Africans suffered from the same disease, the medical perception varied significantly. Third, racial segregation performed itself on a metaphorical level. Though doctors were aware of the infectious nature of sleeping sickness, they conceived of it as an African disease that was not only tied to the continent but also to its people. This was particularly true where the doctors and administrations encountered problems that undermined medical visions of large-scale solutions. The more colonial planning failed, the more Africans were held responsible for the enduring epidemic. This shift impacted on the relationship between both groups: more and more doctors tended to see Africans as the source of epidemic and Europeans as potential victims.

This chapter has aimed to highlight and explain the close ties between colonial rule and medical thinking. When fighting the sleeping sickness became a colonial project, racial categories were widely employed in the everyday process of tropical medicine. Perceiving Africans as a collective and, along with that, as part of nature led to dehumanizing images of African patients. Recent scholarship on disturbing medical drug trials enacted during sleeping sickness campaigns, with devastating effects on African patients, shed light on the outcomes of these perceptions (Neill 2008). To be sure, colonial doctors treating infected Africans not as patients but as research subjects points to the underlying racist basis of colonialism. The aim of this chapter was to show that medical thinking in racial categories did not come out of the blue but was maintained and shaped in daily medical routines. With respect to sleeping sickness, it produced the narrative of African collective culpability. With respect to global public health measures, it draws attention to the colonial genesis of practices in epidemiology and tropical medicine.

Sarah Ehlers is a research fellow at the University of Leicester (United Kingdom). After studying history, political science and cultural studies in Aix en Provence, Zurich and Berlin, she completed her doctoral studies with a dissertation on colonial sleeping sickness campaigns and their impact on European medicine at Humboldt University Berlin (Germany).

Notes

1. On Churchill, see Emmert (1989). On scientific knowledge and colonialism, see Tilley (2011).
2. See for example Driver and Martins (2005) and Pratt (1992).
3. For an analysis of racial thinking in medical demography in colonial Angola, see Samuël Coghe's chapter in this volume.
4. On the early sleeping sickness research, see Soff (1969) and Davies (1962a, 1962b).
5. For a discussion of the causation of the epidemics, see Fevre et al. (2004: 567–573). Lyons (1992) and Ford (1971) consider the disruptive effects of colonization to be responsible for the epidemics.
6. On British Africa, see Hoppe (2003); on the Belgian Congo, see Lyons (1992); on French Africa, see Bado (1996) and Headrick (1994); on German Africa, see Isobe (2009) and Eckart (1997); for a comparative approach, see Worboys (1994); for a comparison of German and French approaches to the disease, see Neill (2012).
7. See, for example, Sibeud (2012), Harrison (2005) and Vaughan (1991).
8. 'Probing the Mysterious Sleeping Sickness', *The New York Times*, 19 April 1908.
9. For similar depictions, see, for example, Hoffmann (1910: 232) and Koch (1912d: 574).
10. On raising awareness for health care, see Meixner 1910; on 'European Needs', see Ziemann (1912: 139).
11. See also Ziemann (1907). The link between lack of hygiene and sleeping sickness is depicted as widespread argumentation by Lyons (1992: 107).
12. Patrick Manson stated that fever occurred only in infected Europeans (Manson 1905: 124).
13. See, for example, the Christy's reports for the Royal Society's Commission on Sleeping Sickness, e.g. Christy (1903). See also Anonymous (1903: 15; 1904: 12). This routine was still common in the 1920s, see, for example, Anonymous 1926–1927; for a French example, see Hubert (1906); for a German example, see Feldmann (1908a).
14. Patient identification with numbered tags was introduced by Robert Koch's team on Sese islands in 1906 (Koch 1912d: picture 19, p. 579; Kleine 1949: 41). See also the picture entitled 'A sure way to see that the patient gets its proper dose is to paint the amount on his dark skin with chalk' from a Cameroun sleeping sickness camp (Gregory and Raven 1936: 528).
15. For numerous camp reports, see the volumes of the Sleeping Sickness Bureau Bulletin (1908–1912) and Medizinal-Berichte über die Deutschen Schutzgebiete, volumes 1907–1913, edited by Reichs-Kolonialamt, Colonial Reports: Uganda, volumes 1906–1914.
16. As synonyms in Scheube (1900: 561).
17. For 'racial immunity' to sleeping sickness, see Rabinowitsch and Kempner (1903); Martini calls sleeping sickness 'this negro disease' (1903: 502).
18. See, for example, Sleeping Sickness Bureau (1910), ASTH (1909), Maier (1909). See also five reviews of German and French cases: Sleeping Sickness Bureau (1912).
19. See, for example, Koch (1912f: 527–528), Manson (1908), Sleeping Sickness Bureau (1910), Daniels (1911), Ortholan (1911), Martin and Darré (1908).
20. On building a separated medical structure in Uganda, see Plehn (1902), Boahen (2011: 105–107); on Togo, see Sebald (1988: 508–509); on Cameroon, see Neill (2012: 124); on French Equatorial Africa, see Headrick (1994: 92–93).
21. See, for example, Schmidt (1910: 408) and Ziemann (1907). See also Grosse (2003) and Stoler (1989).

22. On caged river transport of African workers in FEA, see Headrick (1994: 75ff.); for the Belgian Congo, see Lyons (1992: 108), Arnold (1988: 108). This technique was also mentioned at the international sleeping sickness conference in London 1907/08 (Neill 2012: 124).
23. See, for example, Sambon (1904: 62), Jeanselme and Rist (1909: 106), Dutton and Todd (1905: 90), Martin and Lebœuf (1908: 381–383).
24. See Crozier (2007: 396–398).
25. See also Gradmann (2005: 334).
26. For a detailed description of mass screenings, see Lyons (1992: 84–88).
27. On implementing segregation policies in Douala and Brazzaville, see Neill (2012: 73–79).
28. For example, Kermorgant 1908: 173. See also Headrick (1988: 145–147); on medical influences on colonial town planning, see Ngalamulume (2012: 16–49).
29. See the map 'Schlafkrankenlager in Ajoshöhe. Situationsplan' in Bauche (2007: 10).
30. The London Sleeping Sickness Bureau published in 1910 a collection of all documented sleeping sickness infections of Europeans and counted fifty cases (277).
31. Travel restrictions for Africans were also a common measure in the campaigns against typhus in South Africa, see Marks and Anderson (1988: 274–275).
32. See Koch (1912c: 544–546), Bassenge (1909); see also Tilley (2004: 29–30) and Hoppe (2003: 14–17).
33. For similar procedures in the Belgian Congo, see Lyons (1985: 73-87).
34. On the discussion, see Neill (2012: 38, 158).
35. There are numerous examples in GStA/PK, I. HA, Rep. 76, VIII B, No. 4119; on problems in French infected areas reports, see, for example, Gouzien (1908: 29–71, 64–67).
36. Report of Dr Neubert, Ujiji, 1 May 1906, GStA/PK, I. HA, Rep. 76, VIII B, No. 4118, p. 144.
37. Ibid.: No. 4119, p. 90, Discussion of the German Reichs-Gesundheitsrat, Subcommittee Schlafkrankheit, 10 December 1907; further examples in ibid., No. 4118–4121.
38. See Macleod (2000), Tilley (2011: 196–216). See also Jean Mitchell's chapter on environmental campaigns against malaria in this volume.
39. See Worboys (1994: 91), Rogers (1989).
40. See Packard (2007: 118–122).
41. See Yorke (1913), Bagshawe (1911: 5–7) and Koch (1912b).
42. See Tilley (2011: 191–209), Hoppe (2003) and Isobe (2009: 118–151).
43. Many more examples of blaming Africans for failing tsetse control schemes can be found in *Medizinal-Berichte über die Deutschen Schutzgebiete*, volumes 1907–1915, edited by Reichs-Kolonialamt, Colonial Reports: Uganda, volumes 1906–1914.

Bibliography

Anonymous. 1903. 'General Report on the Uganda Protectorate for the Year ending March 31, 1903', *Colonial Reports – Africa* 15.

Anonymous. 1904. 'General Report on the Uganda Protectorate for the Year ending March 31, 1904', *Colonial Reports – Africa* 12.

Anonymous. 1908. 'Probing the Mysterious Sleeping Sickness', *The New York Times*.

Anonymous. 1908. 'Professor Koch's View on the Problem of Sleeping Sickness', *Journal for Tropical Medicine* 11: 55–57.

Anonymous. 1909. Maßregeln für Europäer zum Schutze gegen die Schlafkrankheit, 25 November 1909, in *Bundesarchiv Berlin* (hereafter: BAB), R 1001/5876.

Anonymous. 1926–1927. Report on a Sleeping Sickness Investigation in Kenya, 1926–1927, in National Archives of Britain (hereafter: NA) CO, 533/379/12.

Arnold, D. (ed.) 1988. *Imperial Medicine and Indigenous Societies*. Manchester: Manchester University Press.

ASTH. 1904. 'Rezension von Manson, Patrick: Sleeping Sickness and Trypanosomiasis in a European. Death: Preliminary Note', *ASTH* 8: 137–138.

—— 1909. 'Rezension von Martin et Guillain. Un cas de trypananosomiase chez un Européen. La forme médullaire de la maladie du sommeil', *ASTH* 13: 101.

Bado, J.P. 1996. *Médecine coloniale et grandes endémies en Afrique 1900-1960. Lèpre, trypanosomiase humaine et onchocercose*. Paris: Éditions Karthala.

Bagshawe, A.G. 1911. 'Recent Advances in our Knowledge of Sleeping Sickness No. III', *Transactions of the Society of Tropical Medicine and Hygiene* 5: 1–37.

Bassenge, R. 1909. 'Rezension von "Feldmann, 'Die Schlafkrankheit im Bezirk Schirati', in Deutsche Medizinische Wochenschrift, No. 14, 1908"', *ASTH* 13: 188–189.

Bauche, M. 2007. 'Trypanosomen und Tinbeef. Medizinisches Wissen um Schlafkrankheit zwischen Kamerun und Deutschland, 1910-1914', *Beiträge zur 1. Kölner Afrikawissenschaftlichen Nachwuchstagung (KANT I)*.

Beck, M., Prof Dr 1909. Report, 25 November 1909, BAB, R 1001/5876.

Boahen, A.A. 2011. *African Perspectives on European Colonialism*. New York: Diasporic Africa Press.

Bruce, D. 1908. 'Sleeping-Sickness in Africa', *Journal of the Royal African Society* 7: 249–260.

Christy, C. 1903. 'The Epidemiology and Etiology of Sleeping Sickness in Equatorial East Africa, with Clinical Observations, 14.07.1903', *Reports of the Sleeping Sickness Commission – Royal Society* 3: 3–32.

Churchill, W. 1989 [1908]. *My African Journey*. London: Cooper.

Crozier, A. 2007. 'Sensationalising Africa: British Medical Impressions of Sub-Saharan Africa, 1890–1939', *The Journal of Imperial and Commonwealth History* 35: 393–415.

Daniels, C.W. 1911. 'Trypanosomiasis in Whites', *Journal of the London School for Tropical Medicine* 1: 67–80.

Davies, J. 1962a. 'The Cause of Sleeping Sickness? Entebbe 1902-03. Part I', *The East African Medical Journal* 39: 81–99.

—— 1962b. 'The Cause of Sleeping Sickness? Part II', *The East African Medical Journal* 39: 145–160.

Driver, F. and L. Martins (eds). 2005. *Tropical Visions in an Age of Empire*. Chicago and London: University of Chicago Press.

Dutton, J.E., and J.L. Todd. 1905. 'Human Trypanosomiasis and its Relation to Congo Sleeping Sickness', *Journal for Tropical Medicine* 8: 90.

Dutton, J.E., J.L. Todd and C. Christy. 1905. 'Two Cases of Trypanosomiasis in Europeans', *Journal for Tropical Medicine* 8: 91–93.

Eckart, W.U. 1997. *Medizin und Kolonialimperialismus. Deutschland, 1884-1945*. Paderborn: Schöningh.

Emmert, K.R. 1989. *Winston S. Churchill on Empire*. Durham: Carolina Academic Press and the Claremont Institute for the Study of Statesmanship and Political Philosophy.

Feldmann, O.. 1908a. Report, 20 May 1908, in BA B R 1001 5898.

—— 1908b. Letter to Gouvernement Cameroon, 18 December 1908, GStA/PK, I. HA, Rep. 76, VIII B, No. 4120.

Fevre, E.M., P.G. Coleman, S.C. Welburn and I. Maudlin. 2004. 'Reanalyzing the 1900-1920 Sleeping Sickness Epidemic in Uganda', *Emerging Infectious Diseases* 10: 567–573.

Ford, J. 1971. *The Role of the Trypanosomiases in African Ecology. A Study of the Tsetse Fly Problem*. Oxford: Clarendon Press.

Fülleborn, F. 1904. 'Bericht über eine Reise zum Besuch des Kongresses der British Medical Association zu Oxford'. Letter to the Prussian Ministry of Health, Education, and Culture 1 August 1904. Geheimes Staatsarchiv Preußischer Kulturbesitz (hereafter: GStA PK), I. HA Rep. 76 Kultusministerium, VIII B, Nr. 4117.

Gouzien, P. 1908. 'La maladie du sommeil dans le Haut-Sénégal et Niger', *AHMC* 11: 29–71.

Governor of German East Africa. 1908. Letter to Ministry of Culture, 4 January 1908, GStA/PK, I. HA, Rep. 76, VIII B, No. 4119.

Gradmann, C. 2005. *Krankheit im Labor. Robert Koch und die medizinische Bakteriologie*. Göttingen: Wallstein.

Gregory, W.K. and H.C. Raven. 1936. 'In Quest of Gorillas: XIII. Gorillas, Men and Sleeping Sickness', *The Scientific Monthly* 43: 522–540.

Grosse, P. 2003. 'Turning Native? Anthropology, German Colonialism, and the Paradoxes of the „Acclimatization Question", 1885-1914', in H.G. Penny and M. Bunzl (eds), *Worldly Provincialism: German Anthropology in the Age of Empire*. Ann Arbor: University of Michigan Press, pp. 179–197.

Harrison, M. 2005. 'Science and the British Empire', *Isis* 96: 56–63.

Headrick, D.R. 1988. *The Tentacles of Progress. Technology Transfer in the Age of Imperialism, 1850-1940*. New York: Oxford University Press.

Headrick, R. 1994. *Colonialism, Health and Illness in French Equatorial Africa. 1885 - 1935*. Atlanta: African Studies Association Press.

Hearsay, H. 1910. 'Sleeping Sickness. Diary, Part IX', *Journal for Tropical Medicine* 13: 167–169.

Hoffmann. 1910. 'Die Aetiologie der Schlafkrankheit', *Verhandlungen des Deutschen Kolonialkongresses Berlin, Verlag Kolonialkriegerdank*, 199–233.

Hoppe, K.A. 2003. *Lords of the Fly. Sleeping Sickness Control in British East Africa, 1900-1960*. Westport: Praeger.

Hubert, H. 1906. 'Distribution géographique des mouches tsé-tsé au Dahomey', Report, Ministère de la France d'Outre-mer, December 1906, in *Archives Nationales d'Outre-Mer* (hereafter: ANOM), BIB SOM. C/BR/10024.

Isobe, H. 2009. *Medizin und Kolonialgesellschaft. Die Bekämpfung der Schlafkrankheit in den deutschen 'Schutzgebieten' vor dem Ersten Weltkrieg*. Berlin: Lit Verlag.

Jeanselme, E. and E. Rist. 1909. 'Trypanosomiase humaine (Maladie du Sommeil)', in E. Jeanselme and E. Rist (eds), *Précis de Pathologie Exotique*. Paris: Masson, pp. 90–124.

Kermorgant, A. 1908. 'Circulaire au sujet des moyens propres à combattre et à enrayer les trypanosomiases et la maladie du sommeil en particulier en Afrique Équatoriale', *AHMC* 11: 169–176.

Kleine, F.K. 1915. 'Schlafkrankheit. Bericht über die Schlafkrankheitsbekämpfung im Jahre 1911/12', in Reichs-Kolonialamt (ed.), *Medizinal-Berichte über die Deutschen Schutzgebiete für das Jahr 1911/12*. Berlin: Ernst Siegried Mittler und Sohn, pp. 94–100.

____ 1949. *Ein deutscher Tropenarzt*. Hannover: Schmorl & v. Seefeld.

Koch, R. 1902. Letter to Prussian Ministry of Health, Education, and Culture, 12 April 1902 in Koch 1912e (925).

____ 1904. Letter to Prussian Ministry of Health, Education, and Culture, 21 September 1904, GStA/PK, I. HA, Rep. 76, VIII B, No. 4117.

—— 1912a. 'Beratung des Reichgesundheitsrats über den gegenwärtigen Stand der Schlafkrankheit in Deutsch-Ostafrika und ihre Bekämpfung, am 05. April 1909', in J. Schwalbe (ed.), *Gesammelte Werke von Robert Koch, Band 2.* Leipzig, pp. 945–949.

—— 1912b. 'Bericht über die Tätigkeit der zur Erforschung der Schlafkrankheit im Jahre 1906/07 nach Ostafrika entsandten Kommission (1909)', in J. Schwalbe (ed.), *Gesammelte Werke von Robert Koch, Band 2.* Leipzig, pp. 582–645.

—— 1912c. 'Schlußbericht über die Tätigkeit der deutschen Expedition zur Erforschung der Schlafkrankheit', in J. Schwalbe (ed.), *Gesammelte Werke von Robert Koch, Band 2.* Leipzig, pp. 534–546.

—— 1912d. 'Über meine Schlafkrankheits-Expedition. Vortrag, gehalten in der Abteilung Berlin-Charlottenburg der Deutschen Kolonialgesellschaft am 24. Februar 1908', in J. Schwalbe (ed.), *Gesammelte Werke von Robert Koch, Band 2.* Leipzig, pp. 563–581.

—— 1912e. 'Über Schlafkrankheit. Korrespondenz', in J. Schwalbe (ed.), *Gesammelte Werke von Robert Koch, Band 2.* Leipzig, pp. 925–929.

—— 1912f. 'Zweiter Bericht über die Tätigkeit der deutschen Expedition zur Erforschung der Schlafkrankheit (1906)', in J. Schwalbe (ed.), *Gesammelte Werke von Robert Koch, Band 2.* Leipzig, pp. 525–533.

Kudicke. 1907. Bericht über das Schlafkrankenlager Kigarama, 1 October 1907, GStA/PK, I. HA, Rep. 76, VIII B, No. 4119.

Lyons, M. 1985. 'From ‚Death Camps' to Cordon Sanitaire: The Development of Sleeping Sickness Policy in the Uele District of the Belgian Congo, 1903-1914', *The Journal of African History* 26: 69–91.

—— 1992. *The Colonial Disease. A Social History of Sleeping Sickness in Northern Zaire, 1900-1940.* Cambridge – New York: Cambridge University Press.

Macleod, R.M. (ed.) 2000. *Nature and Empire. Science and the Colonial Enterprise.* Chicago: University of Chicago Press.

Maier, M. 1909. 'Rezension zu Manson, Patrick: My Experience of Trypanosomiasis in Europeans and its Treatment by Atoxyl and other drugs, in Annals of Tropical Medicine and Parasitology 2 (1908)', *ASTH* 13: 189.

Manson, P. 1903. 'Sleeping Sickness and Trypanosomiasis in a European: Death: Preliminary Note', *British Medical Journal* 2240: 1461–1462.

—— 1905. 'Trypanosomiasis and Sleeping Sickness', in P. Manson (ed.), *Lectures on Tropical Diseases.* Keener: Chicago, pp. 107–130.

—— 1908. 'My Experience of Trypanosomiasis in Europeans and its Treatment by Atoxyl and Other Drugs', *Annals of Tropical Medicine and Parasitology* 2: 33–51.

—— 1910. 'Excessive Liability of European Women in Africa to Trypanosomiasis: Its Cause and its Prevention', *British Medical Journal* 1(2558): 72.

Marks, S. and N. Anderson. 1988. 'Typhus and Social Control: South Africa, 1917-1950', in R.M. Macleod and M.J. Lewis (eds), *Disease, Medicine, and Empire. Perspectives on Western Medicine and the Experience of European Expansion.* London and New York: Routledge, pp. 257–283.

Martin, G. and M. Lebœuf. 1908. 'Étude clinique sur la trypanosomiase humaine (maladie du sommeil)', *AHMC* 11: 381–393.

Martin, L. and H. Darré. 1908. 'Trypanosomiase chez les blancs', *Bulletin de la Société de pathologie exotique* 1: 569.

Martini, E. 1903. 'Protozoen im Blute der Tropenkolonisten und ihrer Haustiere', *ASTH* 7: 499–506.

Meixner, H. 1910. 'Die Bekämpfung der Schlafkrankheit', *Verhandlungen des Deutschen Kolonialkongresses Berlin, Verlag Kolonialkriegerdank*, 257–271.

Neill, D. 2008. 'Paul Ehrlich's Colonial Connections: Scientific Networks and Sleeping Sickness Drug Therapy Research, 1900-1914', *Social History of Medicine* 22: 61–77.

_____ 2012. *Networks in Tropical Medicine. Internationalism, Colonialism, and the Rise of a Medical Specialty, 1890-1930.* Stanford: Stanford University Press.

Ngalamulume, K.J. 2012. *Colonial Pathologies, Environment, and Western Medicine in Saint-Louis-du-Senegal, 1867-1920.* New York: Peter Lang.

Ortholan. 1911. 'Un cas de trypanosomiase humaine', *Bulletin de la Société de pathologie exotique* 4: 624–626.

Packard, R.M. 2007. *The Making of a Tropical Disease. A Short History of Malaria.* Baltimore: Johns Hopkins University Press.

Plehn, A. 1902. 'Rezension zu "Moffat, R. U., Principal Medical Officer's Report for the Year Ending, 31st December 1900"', *ASTH* 6: 99–101.

Pratt, M.L. 1992. *Imperial Eyes. Travel Writing and Transculturation.* London: Routledge.

Rabinowitsch, L. and W. Kempner. 1903. 'Die Trypanosomen in Mensch- und Tierpathologie, sowie vergleichende Trypanosomenuntersuchungen', *Centralblatt für Bakteriologie, Parasitenkunde und Infektionskrankheiten* 34: 804–822.

Rogers, N. 1989. 'Germs with Legs: Flies, Disease, and the new Public Health', *Bulletin of the History of Medicine* 63: 599–617.

Sambon, L. 1904. 'The Elucidation of Sleeping Sickness', *Journal for Tropical Medicine* 7: 61–63.

Scheube, B. 1900. *Die Krankheiten der warmen Länder. Ein Handbuch für Ärzte.* Jena: G. Fischer.

Schmidt, P. 1910. 'Über die Anpassungsfähigkeit der weißen Rasse an das Tropenklima', *ASTH* 14: 397–416.

Scott, J.C. 1998. *Seeing Like a State. How Certain Schemes to Improve the Human Condition have Failed.* New Haven: Yale University Press.

Sebald, P. 1988. *Togo 1884 - 1914. Eine Geschichte der deutschen ‚Musterkolonie' auf der Grundlage amtlicher Quellen; mit einem Dokumentenanhang.* Studien über Asien, Afrika und Lateinamerika 29. Berlin: Akademie-Verlag.

Sibeud, E. 2012. 'A Useless Colonial Science?', *Current Anthropology* 53(S5): S83-S94.

(London) Sleeping Sickness Bureau. 1909a. 'Influence of External Conditions on the Distribution and Numbers of Glossina Palpalis, Sleeping Sickness Bulletin', *Sleeping Sickness Bureau Bulletin* 1: 94–106.

_____ 1909b. 'Personal Prophylaxis of Sleeping Sickness', *Sleeping Sickness Bureau Bulletin* 1: 486.

_____ 1909c. 'The Prophylaxis of Sleeping Sickness', *Sleeping Sickness Bureau Bulletin* 1: 106–118.

_____ 1909d. 'Sleeping Sickness: How to Avoid Infection', GStA/PK, I. HA, Rep. 76, VIII B, No. 4120.

_____ 1910. 'Sleeping Sickness in Whites', *Sleeping Sickness Bureau Bulletin* 2: 277–321.

_____ 1912. 'Trypanosomiasis in Whites', *Sleeping Sickness Bureau Bulletin* 4: 18–25.

Soff, H.G. 1969. 'Sleeping Sickness in the Lake Victoria Region of British East Africa, 1900-1915', *African Historical Studies* 2: 255–268.

Stoler, A.L. 1989. 'Making Empire Respectable: The Politics of Race and Sexual Morality in 20th-Century Colonial Cultures', *American Ethnologist* 16: 634–660.

Tilley, H. 2004. 'Ecologies of Complexity: Tropical Environments, African Trypanosomiasis, and the Science of Disease Control in British Colonial Africa, 1900-1940', *Osiris* 19: 21–38.

_____ 2011. *Africa as a Living Laboratory: Empire, Development, and the Problem of Scientific Knowledge, 1870-1950.* Chicago: University of Chicago Press.

Vaughan, M. 1991. *Curing their Ills. Colonial Power and African Illness*. Stanford: Stanford University Press.

WHO. 2006. 'African Trypanosomiasis (Sleeping Sickness)', Fact Sheet 259. Retrieved 20 July 2014 from: http://www.who.int/mediacentre/ factsheets/fs259/en.

Worboys, M. 1994. 'The Comparative History of Sleeping Sickness in East and Central Africa, 1900-1914', *History of Science* 32: 89–102.

Yorke, W. 1913. 'The Relation of Big Game to Sleeping Sickness', *Journal of the Royal African Society* 13: 23–32.

Ziemann, H. 1907. *Wie erobert man Afrika für die weiße und die farbige Rasse*. Barth: Leipzig.

——— 1912. 'Über die Schlafkrankheit in Großkamerun', *Beihefte zum Archiv für Schiffs- und Tropenhygiene* 16: 112–140.

5
Colonial Histories of Cancers
Primary Liver Cancer in Africa, 1900s–1960s

Jean-Paul Bado

Most historical studies of twentieth-century colonial medicine in Africa have dealt with epidemic and endemic diseases. They generally addressed health with respect to a colony or a region and all of them established reasons for reduced life expectancy. In European discourses, Africa has often been viewed as the continent of diseases, 'a White man's grave', a place where Africans suffered from many ailments and from malnutrition which impacted their health and their immunity. It was the continent that certain colonial physicians turned to in an attempt to understand how Charles Robert Darwin's evolution theories (1859) would apply to humans.

I began to wonder about cancer while reading historian Pierre Darmon's book (1993) and that of biologist Patrice Pinell (1992), along with the *Archives de médecine navales*, the *Bulletin de la Société de Pathologie exotique*, the *Annales d'hygiène et de medicine coloniales*, the *Annales de l'Institut Pasteur*, etc. I remembered the pyramid diagram representing African populations by age group in the French colonial empire, and I noted what many others have observed: the pyramid base was broad and wide, its peak narrow and pointed. The usual conclusion is that populations will not live long enough to develop cancers because cancer only develops in populations of advanced age. I wondered about the role of biomedicine in promoting this idea of the cancer-free African.

Emile Marchoux, a colonial physician who founded the first Pasteur Institute in Saint-Louis (Senegal) in 1896, explained that the conquest of Africa could begin when European physicians were able to vanquish its diseases (Bado and Michel 1991). Many of these diseases were infectious diseases or were linked to the environment, but colonial physicians did not include cancer, which was, even in Europe, a neglected disease before the 1920s. Some colonial physicians believed that cancer was an unknown affliction in Africa. They knew its symptoms but not its causes or its pathogenic germs, because cancer epidemiology did not respect the scientific model of knowledge proposed by Louis Pasteur, Robert Koch, and other famous scholars, with the exception of Rudolf Virchow.

This ailment provoked considerable debates among modern medical specialists in Europe, in the USA, in Japan, and in Africa. In fact, many scientists wondered why certain diseases affected only 'Black' or 'Yellow' people and not every human being; at that time human beings were described according to their 'skin colour'; hence, 'Black', 'White', 'Yellow', 'Red Skin', etc., or, to make it even more complicated, 'half-caste'. 'Black Africa' was used to characterize Sub-Saharan Africa, despite the presence of 'White people' in South Africa, Madagascar and Ethiopia, and that of the Tuareg and Fulani in the Sahel region. Other physicians, however, expressed doubts about race-specific diseases. Laboratory research and biomedicine seemed to provide the right kind of approach for understanding these issues thanks to advancing research techniques. The Pasteur Institute became one of the global leaders in this regard.

In our respective books, Helen Tilley (2011) and I (1996) highlight the lack of biomedical knowledge, and portray colonial doctors in Africa as curious as to why some populations survived diseases that seemed to be lethal to other groups. Along the same lines, I now ask why some colonial physicians denied cases of cancer among Africans, more particularly in groups they termed 'Black Africans',[1] as they had already for malaria and yellow fever. Megan Vaughan (1991), myself (1993), and other authors (Scott 1965; Headrick 1994) have highlighted that during the nineteenth and twentieth centuries, colonial physicians frequently identified the African 'lifestyle', that included cultural and environmental elements, in their explanations for health and illness. Furthermore, Vaughan and I have shown that biomedical discourses related to Africa suffered from internal contradictions due to physicians' ignorance about disease and about the African 'lifestyle'. We can summarize physicians' goals throughout the twentieth century using three verbs: to observe, to diagnose, and to heal. Colonial medicine and biomedicine continued to change thanks to new knowledge that led to the abandonment of erroneous theories. As Pierre Thuillier noted in 1980, '"science" is not what some call the methodical and disinterested quest for knowledge; it is a force that manifests itself more and more visibly in all sectors of life' (Thuillier 1980).

I seek to illustrate the slow changes of colonial medicine and biomedicine specifically with respect to primary liver cancer, because of the growing understanding of the liver's crucial role in maintaining health through the production of enzymes and bile. Also, this paper explores the debates on the very existence and etiologies of cancer, particularly primary liver cancer, in late colonial Africa. Some specialists of colonial medicine embraced the myth of the *bon sauvage*, systematically denying that colonized African people could even suffer from cancer, because cancer was predominantly a disease of civilization – at least European civilization. Other cancer specialists disagreed with these claims and disputed both the methods and evidence used to deny the existence of cancer in Africa. For these researchers, histological tests suffice to confirm the existence of cancer among Africans. Histological tests, however, could not explain what provoked the change in the cells. Through an historical analysis of debates regarding the existence and causes of cancer, I reveal biomedicine to be a hostage of its older colonial ideology (Monnier 1999) despite earnest adherence to scientific rigour (Kermorgant 1901; Peyrot 1905; Jambon 1914; Spire 1919–1920; Bado 1999). Through this analysis, we will encounter the geographical distributions of primary liver cancer, the environmental exposure that cause it, along with measures undertaken to prevent its incidence. The discovery of micotoxin and aflatoxin contributed to blaming African environments without resolving the problem of the origin of primary liver cancer. After decades of debate, thanks to a vaccine against hepatitis B, it became possible to prevent primary liver cancer in Africa and other continents.

A History of Knowledge about Cancer and Primary Liver Cancer

When I discovered the problem of cancer, in particular primary liver cancer (PLC), in an autopsy report (Calmet and Journe 1946), I realized that I had to revise my interpretation of population pyramids in relation to cancer. This report about one of the *tirailleurs sénégalais* (colonial soldiers originating from French colonies of Africa) from 1946 was so astonishing that I was initially sceptical that cancer was the cause of his death. The soldier from French colonial Guinea who died of primary liver cancer (PLC) in Indochina was a practicing Muslim who did not drink alcohol. The forensic surgeon concluded in his report that the cause of death was PLC and pointed out that many such cases had been detected in Africa. Yet, at that time, many cancer specialists attributed PLC to cirrhosis – the result of alcoholism (ibid.).

In the mid-twentieth century, many illnesses contributed to adult mortality in Africa, PLC being one of the more common ones. From the autopsy reports by colonial physicians, written from the early 1930s onwards (Rongier 1931), we know today that PLC could strike young adults in the prime of their lives. Its incubation period could last for decades, appearing when the patient was between thirty and forty years old, sometimes earlier and sometimes later. Even nowadays, cancer, especially PLC, affects population cohorts between thirty and forty years old, contributing to the overall mortality and to the shape of age pyramids in Africa. Cancer has contributed and continues to contribute to the low percentage of old people in the population living between the Tropics of Cancer and Capricorn, and beyond. Hence, the African age pyramid cannot be viewed as an explanation for the seeming absence of cancer in Africa; instead, cancer needs to be considered as a causal factor that shapes the age pyramid.

By the early twentieth century, cancer was considered by many colonial physicians and European cancer specialists as a 'disease of civilization' afflicting mainly Europeans or others from Western (developed) countries. The specialists therefore believed that this disease rarely occurred in Africa.

In 1909, Ortholan, a physician to French colonial troops, wrote an informative essay related to cancers in the tropical world. Notably, he was not only referring to the regions between the Tropic of Cancer and Tropic of Capricorn, but also to other regions where it is possible to find warm climates, as Patrick Manson (1898) explained. Ortholan (1909) wondered if cancers were, in general, as frequent in the tropical world as in temperate countries. He claimed in published studies that he found no mention of cancers or malignant neoplasms (often called tumours) in Africa, although he acknowledged that perhaps physicians there paid no attention to the pathologies, or that they had not investigated the 'native' population sufficiently.

Despite insufficient research concerning the absence of tumours among 'black people', as well as their frequency and their causes in Europe, Ortholan hastily concluded that cancer occurrence was rare in 'Black Africa' (ibid.:). In fact, he drew on the work of Rufz de Lavison (1869) in Martinique from 1837 to 1856. After observing ninety-three cases of cancer, Rufz de Lavison contended that the number of cases was very low. Among thirty-three uterine cancers, he underlined that only three were diagnosed in 'black women'. The frequency of uterine polyps in Martinique led him to believe that there was some special element in the blood of 'black people' which produced fibrous bodies (ibid.), and that 'the Blacks' had in their organisms what was necessary to protect themselves against cancer (ibid.).

In spite of these errors, many well-known practitioners, including Ortholan, concurred with Rufz de Lavison's findings. However, medical scientists observing the occurrence of PLC in Dakar's Indigenous hospital in 1920 insisted that PLC did indeed exist in Africa.

In 1939 a medical scientist, Robert Dupont, who undertook a retrospective study on the cancer question in French Africa (from Algeria to equatorial Africa), was highly critical of earlier studies mentioned above by Mouchet and Gérard (1919, 1926), Adam (1924), Nogue (1920) and Cazanove (1931) on PLC. These studies underlined the presence of cancer and of PLC in particular. Afterwards, in 1942, Dupont published his own medical observations related to cancer. Despite his efforts to identify cases of cancer, he admitted that he had found none among the thousands of sick persons examined (Dupont 1942). He was convinced that cancer was absent (ibid.), even though he felt obliged to report 156 cases discovered in 1939 in French West Africa (AOF) [Afrique Occidentale Française] (Dumas 1939), along with fifty-one cases in Cameroon (Bablet 1935) and sixty-two in Belgian Congo (idem 1947). Dupont firmly denied the presence of cancer in Chad (1942). This claim would, however, be refuted by Léon Palès, who had diagnosed many cases in that territory (1946).

Notably, Dupont rejected the claim of other physicians that cancer did not develop among 'black people' because they did not reach 'cancer age'. According to him, the absence of cancer, or at least its rarity among 'the Blacks', was due to their diet: little meat but a lot of millet that is rich in magnesium. His analysis reinforced Pierre Delbet's studies. Delbet, a cancer specialist, had claimed that cancer was provoked by the lack of magnesium in the human organism (1936). According to Dupont's research on the inhabitants of Chad, the diet of the Sara ethnic group was based on millet very rich in magnesium chloride. This high consumption of magnesium supposedly explained the lack of cancer among them.

Obviously, Dupont's objective was to link the presence of cancer to civilization, as he eventually admitted:

> The populations that we have examined are from a less advanced level of civilization and our observations confirmed the view of some authors that cancer is an ailment of civilization (or at least what we call civilization). What is certain is that it is not a question of race because cancer exists among the 'Black' in America and it exists also in Dakar where it is supposedly as common as in Europe ... In the United States and in Dakar, Blacks have the same diet as the Whites around them. (Dupont 1942: 733; author's translation).[2]

Dupont concluded by suggesting that there was a link between cancer and diet.

Ortholan and Dupont especially botched their investigations by relying on rapid surveys informed by racial prejudices with reference to skin colour and civilization. They based their claims on fragmented data, and they extrapolated their conclusions to different locales, for they could only comment on the colonial towns that had been visited by investigators. Dupont's motives were moreover highly questionable as his investigation was informed by the need to demonstrate that cancer did not exist in African populations.

With his publication, Dupont supported Delbet's claims regarding lifestyles. In his 1936 essay 'Le cancer dans la race noire', Delbert had referred to Frederick Hoffman who had declared that cancer was not necessarily a universal disease but became widespread as a result of the habits of modern life. In Delbet's eyes the African continent had long been the exception, as it was still sheltered from modern influence. According to him, in Africa,

> the main diet of Blacks living in a primitive state in the Ivory Coast is extremely rich in magnesium. [...] The primitives [...] absorb all the magnesium in their diet, which is much richer in magnesium than that of the civilized [people]. [...] I find here the condition I have so often tried to draw your attention to: abundance of magnesium, rarity of cancer, deficiency of magnesium, frequency of cancer. It is possible that are other causes for the increase in cancer frequency among the Black in Africa when they switched from primitive to civilized life. But the deficiency in magnesium appears to me the important factor of this deplorable transformation. (Delbet 1936)

Rejecting the value of the clinical diagnostics of his colleagues in Africa, and pointing to the lack of histological proof by microscopical anatomy, Dupont was in a strong position to impose his conclusions. But Léon Palès (1946) questioned the validity of the methods that had been used to verify the absence of cancer. In 1946, Palès analysed how practitioners used statistical data in their conclusions to confirm or to invalidate the presence of cancer, and underlined the fact that the collection and the utilization of statistics varied because the documents did not come from the same sources, and that many practitioners had taken them literally.

Interestingly, the debate about the existence or inexistence of cancer among the indigenous populations had begun much earlier in the northern part of Africa. After the conquest of Algeria in 1830, in their dermatological works, military physicians had pointed to the presence of cancerous tumours among Algerians, but without providing proof from the perspective of anatomo-pathology (the anatomical aspect of pathology). The affirmation of colonial principles led to a radical change in medical discourse and the denial of cancer for Muslim and Jewish populations. Some of them created an artificial difference between European and local diseases by forgetting that some ailment like smallpox was provoked by the same virus. Yet Jules Brault's 1908 publication of histological tests regarding forty-five cases from his eighteen years of service in Mustapha town (now Sidi M'Hamed in Alger) made it difficult to continue to deny the presence of cancer in Africa.

Partial Data on Cancer and PLC and Medical Contradictions

In French West Africa

From 1919 to 1922, Aristide le Dantec, former director of Dakar's Medical School for Africans, and an ardent advocate of the connection between civilization and the occurrence of cancer, underlined the rarity of cancers in 'Black Africans' (1929). Of the eighteen cases recognized among the patients at Dakar hospital between 1921 and 1926, ten suffered from PLC, two from cancer of the womb, one from an oesophagus cancer and another one from stomach cancer, which proved that there were different types of cancer among 'the Blacks', and that cancer had long been present in Africa. And yet, for le Dantec, the most important thing was to insist on the rarity of cancer among Africans.

But other physicians disagreed, and suggested that cancer was in fact not so rare. A study quoted by Ferdinand Heckenroth and Bergonier (1923) reported that out of 785 deaths in Dakar during 1922, eight were caused by cancer. The identified cases perplexed the specialists as they had mainly occurred in villages not exposed to European contact. In their essay from 1926, Mouchet and Gérard warned physicians who travelled in Africa against the illusion of the extreme rarity of 'neoplasm': 'If among the Black females who consult Europeans there is no carrier of mammary neoplasm, this does not mean that this affliction does not exist. It is simply that these poor women prefer hiding rather than to expose their infirmity' (Mouchet and Gérard 1926). Ferdinand Heckenroth and Bergonier (1923) insisted that they had seen a case of PLC in a thirty-five-year-old Wolof man originating from Tivaouane in Senegal. This man was non-alcoholic and had never left his village. Significantly, this case proved the presence of autochthonous cases of cancer, cases not induced by a European lifestyle that some physicians called civilization.

Other observations of cancer were made in 1926. D'Arusmont's observations of cancer in 1926 (1930) challenged le Dantec's histological assertion. British physicians recognized cases of PLC and other cancers in Sierra Leone, the Gold Coast and Nigeria, which contradicted the conclusions of those in South Africa, Rhodesia (now Zimbabwe), Kenya and Tanganyika concerning the rarity of cancer among indigenous people, despite histological data. Moreover, E. Smith's and B.G. Elmes' 1934 inquiry showed that over a period of eight years, 500 malignant tumours were identified in Nigeria among 'the Blacks' (Adler and Cumming 1923; Smith and Elmes 1934). All these findings made it difficult to continue to deny the existence of cancer, and yet, some physicians did.

It is interesting to note the change in medical knowledge. Biomedicine, through le Dantec, Dupont and others, was reluctant to abandon colonial prejudices. The more the proof accumulated, the more they relied on their prejudices to resist changing their explanations. However, this resistance increasingly became untenable in the face of the pressure of changes in scientific knowledge. Colonial medicine had first denied the presence of cancer in Africa by underestimating cases, and by insisting on the lack of histological tests. Despite the accumulation of local cases, colonial medicine rejected this evidence, the bias of this position becoming increasingly obvious in the face of changes in medical science. J.M. Montpellier and J. Montpellier, drawing from this so-called evidence, concluded that 'all cancer victims were Negroes who had lived a primitive life' (Montpellier and Montpellier 1947). By linking 'Negroes' with 'primitive life', they responded to assertions that continued to link African cancers with colonization.

Beyond Senegal and French Sudan, where connections between the European presence and cancer were increasingly tendentious, Charles Commes and Henri de Vallendé (1916) objected to the 'general belief' regarding the rareness of cancers in tropical Africa. In 1936, the Pasteur Institute in Dakar received sixty-three tumour samples from inhabitants of Senegal and of Sudan, and identified thirty-three malignant tumours, twenty-seven epitheliomas (a malignant growth containing epithelial cell, epithelial cancer), and six sarcomas (tumours usually arising from connective tissue, most of which are malignant). One of these cases was PLC (others occurred in the breast, uterus, stomach, bowels, salivary glands and skin). However, their research was not sufficient to make cancer, particularly PLC, a cause that would mobilize colonial health authorities. Instead, cancer only attracted the interest of two groups of physicians. The first group claimed that cancer was an illness of civilization. The other group sought to show that cancer, like many other diseases, knew no boundaries.

In equatorial Africa and elsewhere

In French Equatorial Africa (FEA) [Afrique Equatoriale Française] and Cameroon, many scholars' studies denied Ortholan's 1909 conclusions. In 1915, J. Ringenbach and Guyomarch (1915) contributed to dismantling Ortholan's exaggerated conclusions. In 1922, Charles Jojot and Jean Laigret (1922) identified a case of fuso-cellular sarcoma in a thirty-year-old Cameroonian. Dupont appears not to have been familiar with this research in which Jojot and Laigret concluded that the 'black race is like other the victim of cancer and more particularly, perhaps, of sarcoma' (ibid.).

Nor did Dupont know of Hansemann's 1914 study that underlined all the varieties of tumours (Montpellier and Montpellier 1947). In 1921, in Fernando Po, Appel, a physician to the German colonial troops, detected 623 deaths including 41 cases of cancer out of 5,000 sick persons (Palès 1946, 1954). In 1938, the physicians of French colonial troops, D'Anella and Tourenc (1938), recognized that cancer should not be considered an exceptional disease.

Not a single indication was provided in the colony of Gabon. In Moyen Congo (now the Republic of the Congo) the Pasteur Institute in Brazzaville reported that from 1930 to 1940 there was no description of cancer. But in 1942, one epithelioma and two melanomas in women between forty and forty-five years old were diagnosed. Inquiry into oncology also took place on the Somali coast where in 1923 a report from the French Health Colonial Service indicated that cancer was unknown among indigenous and European residents. In 1945, A. J. Leitner (1945) generated statistical data that documented tumours in this territory since 1932. From March 1939 to October 1943, twenty-four cases were observed, including a lot of PLC and digestive tract cancers.

In 1909, in Madagascar as well as in La Réunion and the Comoro Islands [les îles Comores], Ortholan concluded that 6.69 per cent of the deaths should be imputable to cancers. One statistic from the provinces stated that 1 per cent of the deaths were caused by cancer. But a report of the International Hygiene Office published in 1924 mentioned that cancer 'seems rare' in Madagascar (cited in Montpellier and Montpellier 1947: 45). D'Anella's statistics in 1933 confirmed the rarity of cancers, despite the discovery of two local cases. The Assistance *médicale Indigène* hospital in Tananarive provided data between 1927 and 1936, indicating the presence of cancer: 182 cases out of 48,442 hospitalizations. From 1938 (Moustardier 1937, 1938) the discourse changed: the Malagasy (or Madagascan), like Europeans, could develop cancers. To sum up, in French West Africa and in French East Africa epidemiological prospecting on cancerology (or oncology) was insufficient and was dominated by a kind of competition between those who held the view that Africans could develop cancers and those who developed the linkage between colonization and cancer.

What were the reasons for a theory connecting colonization and cancer? On the one hand, cancer was perceived as a sickness of European civilization, resulting from a lifestyle based on development and progress that the 'Black people' in Africa were far from attaining. By denying their capacity to develop cancers without European intervention, the physicians wanted to enclose them in a 'state of primitiveness'. As Delbet wrote in 1936, 'It is not immunity of race. The Negro is vulnerable to canceration, but he very rarely succumbs when he remains in a savage state'. This return of a primitive state allowed the classification of human beings based not only on their skin colour, but also in relation to the illnesses that they could develop. If proven right, this theory would serve to comfort proponents of superior and inferior races, furnishing them with a new weapon: cancer, an illness of the superior race. It would then be undeniable that modern medicine dominated by colonial theories had not yet penetrated deeply into the African world. At the same time, however, this connection reveals the fear of civilization, of the progress due to the Industrial

Revolution. This fear guided certain physicians in their opposition to European civilization, itself perceived as a kind of a cancer.

Just as Montpellier emphasized, if these facts came to demonstrate that rates of cancer morbidity were relentlessly increasing and were really attributable to the penetration of modern life, 'one should be tempted to admit that it is a forceful argument in favour of exogenous conception' (1947), that is to say, of the virulence of cancer. The assertion of the absence of cancer gave way to other arguments, which referred to a multitude of illnesses that afflicted populations of Overseas France [France d'Outremer]. In fact, the so-called apparition of cancer among people living far from modern civilization accompanied medical expansion, although certain cancers dependent on industrial production could be attributed to the European presence.

Searching for Epidemiological Factors of Primary Liver Cancer and the End of a Concept of Medicine

J.M. Montpellier and J. Montpellier noted the existence of cancer in all territories of Overseas France, insisting that 'everywhere medical penetration had had an opportunity to investigate and carry out continuous activity, cancer was noted' (ibid.). It had been mentioned since 1876 in Algiers, and in 1905 biomedical specialists identified PLC in South Africa and underlined many cases among 'black people'. People affected by PLC did not benefit from specialists' attention, for in French colonial Africa they were more concerned with the medical care of colonial troops and administration. In addition to numerous results collected on the frequency of cancer, there were many comparative studies concerning pathological environments.

PLC finally attracted serious interest in 1944 and 1946 when clinical and anatomical descriptions were carried out in Dakar. It was then that A. Geyer reported that cancers of all kinds afflicted inhabitants of French West Africa (1946). Out of 3,500 anatomic pieces examined at the Pasteur Institute in Dakar after 1937, 713 malignant tumours were diagnosed by histology; 613 of them were '(black) autochthonous'. In reference to this analysis, Henri Jonchère asserted that 'a few years ago, one frequently heard that cancer was relatively rare amongst the subjects of black race. It does not seem that its appearance is recent' (Jonchère 1948).

Notably, PLC was in first place among cancers. From the 1940s, research focused on the specificity of PLC, which was recognized only amongst 'Blacks'. Thus Kennaway (1944) established a comparison between PLC in 'Black Africans' and 'Black Americans' to make it a disease of racial origin. But the studies revealed that its prevalence among 'Black Americans' was not as high as among 'White Americans'. Faced with these results, the scientists who claimed that PLC was a disease unique to 'Blacks' had to abandon their prejudices.

Beyond French Africa, the question of PLC had captured the attention of British physicians in South Africa where many sick people came from the Portuguese colony of Mozambique (Berman 1941, 1951). In summarizing all the African data (in 1951), Charles Berman estimated that among Bantus of the South African Union, of

Mozambique, of Kenya, of the Belgian Congo and of French Equatorial Africa, PLC represented 37.4 per cent of all cancers (ibid.). Among the Blacks of Guinea, Senegal, Sierra Leone and Nigeria, the rate was 18.7 per cent. According to Geyer, the rate of the whole territory of French West Africa was 33 per cent (Berman 1951).

In addition to the clinical and anatomico-pathological aspects, the different authors found that the average age of those affected was between thirty and fifty years, and most of those affected were between thirty-one and forty. In 1951, Guérin calculated the same averages (1952). Surprisingly, too, many children had succumbed to PLC, even a three-day-old baby. Nevertheless, this case did not point to the origins of the illness. Researchers were instead interested in the distribution by gender of the patients, since many patients were men and only a few were women. For many authors this gap came from endogenous factors that were impossible to discover. In 1951, sexual hormones were suspected along with biliary acid. Doctors also suggested that hepatitis carcinogenic factors existed in the liver itself. For Guérin, PLC was provoked by a bad diet (sub-nutrition and malnutrition), multiple parasites, infection with blood and varied hepatitis, etc.

One can understand the conclusions of medical researchers without justifying them. Until 1951, scientists continued to associate PLC with cirrhosis (Bergeret 1950; Pene and Boisson 1975). Afterwards, experimental conclusions, clinical observations and autopsies diminished the ranks of those attributing cancer to cirrhosis. If cirrhosis was the origin of hepatitis cancer, then gaining a better understanding of cirrhosis was a top priority.

Bernard Ninard, at the Pasteur Institute, studied the histogenesis of tumours and in so doing made important progress in cellular theory (1950). He concluded that hepatitis could be at once cirrhogenic and carcinomic, but it could also be one or the other. A sick person could develop cirrhosis without cancer. Guérin approved of Ninard's conclusions, noting that in childhood, from the period of weaning, hepatitis manifested itself frequently and provoked kwashiorkor. Anatomically, kwashiorkor (a syndrome caused by nutrition deficiency and characterized by delayed growth, changes in the skin and hair pigment, etc.) produces alterations in the pancreas and steatosis lesions on the liver. Barbara Cooper studies this disease in her contribution to this volume. The identification of hepatic steatosis (fatty degeneration) from severe childhood malnutrition can thus point to one of the stages of PLC. As Ninard understood for the first time, a child's liver weakens, and is subject to the effects of malaria and of toxic substances that participate in cirrhosis histogenesis (the development of tissue from the undifferentiated germs of cell layers of embryo).

Despite the difficulty in discovering the origin of and difference between PLC, cirrhosis and kwashiorkor, Charles Bergeret concluded that African children at the stage of weaning were not only undernourished, but their diet was so imbalanced that certain food substances were rendered toxic and this caused cirrhosis (1950). M. Payet, R. Camain, and P. Pène noted in 1956 that PLC was almost always associated with cirrhosis, but cirrhosis could not be a secondary reaction of defence; 'the factors of evolution and civilization' prevented its expansion, whereas 'poverty, famine, many infections' (86) favoured it.

In other words, according to Guérin and others, PLC found its origins in a nutritional failure, provoked by the abrupt cessation of breastfeeding, followed by malnutrition. If this was really the case, how might one explain that out of seventy-five cases of cirrhosis in France, physicians only discovered two cases of PLC? How do we explain the case diagnosed in a three-day-old baby in Greece? What indeed explained PLC's frequency in Greece where coroners identified 10.7 per cent cases among 3,828 autopsies, even among breast-feeding infants (Mouchet and Gérard 1926)? For Guérin and others, cirrhosis in 'Blacks' and 'Whites' were not identical (Pène and Boisson 1975). If cirrhosis of 'Blacks' was different from those of 'Whites', one could not speak of cirrhosis in both cases.

Attention then turned to the question of the toxicity of consumed products, notably those of azoic derivatives. Research teams experimented by provoking PLC in mice and rabbits. The macroscopic and histological results were identical to those of human beings. After 1936, many tests in Japan were conducted to determine the role of diet in the origins of PLC, particularly with regard to hydrogenated palm oil. In the 1950s, group B vitamins were suspected of causing PLC. Despite the failure of tests with hydrogenated palm oil, research now focused on African populations' diets, rich in lipids and imbalanced from the point of view of vitamins. Researchers isolated the lipids of palm oils, peanuts, peanut butter, etc., to study the fatty acid suspected of playing a carcinogenic role.

Faced with many failures, researchers decided to progressively exclude different products which were often considered as factors but were absent from regions at high risk. They concentrated studies only on those factors encountered in all regions. They strove to find ecological factors present in all regions concerned, to distinguish the role of food or rations, to assess the action attributed to endemic parasites present in all regions, and to accentuate the research in place on bacterial and viral infections.

The research undertaken in the 1950s came to fruition at the end of the 1960s. Ecological factors, including climate, rainfall, soil composition and geographic situation were definitively set aside. The results of investigations revealed that in identical climates there was no particular frequency of PLC. As for the role of food and nutrition, it was considered quite important. It had been observed for years that the nutritional deficiency which brought about kwashiorkor (Bergeret 1948) was a factor of PLC. Research on patients who had this deficiency due to developing kwashiorkor as children ruled out this nutritional deficiency, since some childhood fatalities had been due to dystrophy associated with parasitic or other infections. Survivors with hepatic steatosis subsequently suffered from nutritional cirrhosis, a 'factor of essential conditioning in the appearance of PLC'. From this, the sequence considered by nutritionists and cancer specialists was: kwashiorkor, steatosis, hepatitis, nutritional cirrhosis, PLC. Notably, Bergeret, in Dakar in 1946, had already insisted on the difficulty in diagnosing adeno-cancer perceived as banal cirrhosis (1950). Hence the problem of disentangling their origin when one considers the superposition of maps concerning cirrhosis and PLC countries (Barret 1975).

After these observations, it was necessary to validate them experimentally so as to identify the carcinogenic factor. Scientists in Great Britain, in France, in the USA and in other countries identified carcinogenic substances originating from consumed plants. Plant products could serve as a substrate (a substance acted upon by an enzyme) for fungal proliferations, and ultimately for the mycotoxins responsible for the artificial contamination of human foods (Darnis 1980; Diop, Denis and Maupas 1981). These were aflatoxins (Peers, Gilman and Linsell 1976; WHO 1980), present in several kinds of food, notably maize, wheat, barley, oats, rye, rice sorghum, peanuts, soy, beats, copra, cottonseeds, and in animal feed. The researchers observed the highest contaminations in peanuts and oil seeds eaten by human beings and animals, which raised many scientific, technical, hygienic and economic questions.

For animals, the aflatoxin M1 contaminated liquid or powdered milk, along with other dairy products. This contamination came from storage modes. The discovery of the turkey X disease in 1960 (Barret 1975), when 100,000 animals died in England, led to intense scrutiny of this toxin. From the investigations conducted to determine the exact cause of this epizootic, scientists observed that turkeys were nourished with flour considerably contaminated with aflatoxins. After identifying this substance through the studies of the Tropical Products Institute in London and those of the Central Veterinary Laboratory in Weybridge, scientists tried to understand the origin of the epizootic by identifying *Aspergillus flavus*. Already in 1945, Ninard and Hintermann had indicated that pigs nourished with oil seeds suffered from hepatitis lesions and tumours, without finding the real cause (1945). In addition, animals such as ferret, rat, trout, salmon and monkey, which consumed aflatoxins (especially substance B1) orally suffered from hepatitis. Nevertheless, each species manifested its own sensitivity to aflatoxin. Aflatoxin was identified as responsible for an epidemic of so-called viral hepatitis among pigs.

With respect to these results, along with reports by Sankalé and others (Barret 1975), aflatoxin came to be recognized as the determining cause in PLC's etiology. But it was not just a question of experimentation. Éliane le Breton (Le Breton 1958: 62–68; Le Breton, Frayssinnet and Boy 1962; Le Breton et al. 1963) had shown in 1963 that white Wistar rats contaminated with aflatoxins developed cancer three months after a cirrhosis period. According to her, a deficiency in choline[3] accelerated the process of canceration and increased the percentage of cancer due to aflatoxins. Thus, researchers had returned to cirrhosis as a precursor to the development of PLC.

Aflatoxin (Diouf 1967) is produced by *Aspergillus flavus*, a toxin that contaminates peanuts and peanut oil cakes – the solid remaining after peanuts are pressed to extract the oil. Oil cakes can be used to feed milk cows because they are so rich in proteins. These proteins attracted the attention of researchers because they could be introduced into human diets. M21 flour (20 per cent cake powder, 70 per cent millet and 10 per cent fish powder) was made by nutritionists in the 1950s to alleviate malnutrition in Africa and elsewhere, and especially malnutrition associated with the abrupt cessation of breastfeeding (ibid.). The implementation of this project should have been undertaken by a large-scale food programme in 1963 (ibid.), under the aegis of United Nations International Children's Emergency Fund (UNICEF) created in December 1946. However, it was at that time that the problems caused by *Aspergillus flavus* were demonstrated, delaying all programmes.

Oil cakes contaminated by the turkey X disesase (the name of this new disease) came from Brazil. Following this contamination, peanut and oil laboratories in Europe, America and Africa led inquiries to demonstrate that the majority of peanuts, wherever they came from, were contaminated. By stopping to consume Oil cake coming from these countries, the X disease disappeared. This announcement was suspect, particularly when counter-investigations confirmed this contamination. The disquiet grew in countries exporting peanuts, especially Senegal, Gambia, Mali, Niger, Nigeria and Sudan, which had heavily invested in the cultivation of peanuts. There followed a series of appraisals and counter-appraisals that allowed the exclusion of sterilized (or pasteurized) oils. Problems with contaminated oil cakes remained since they were used in animal feed and could thus lead to carcinogenic meat, particularly in Western countries which imported oil cakes. In producer countries a panic ensued, with demands that governments should deploy equipment to resolve the crisis.

It was at this moment that specialists intervened to demonstrate that clinical experimentation had nothing to do with nutritional experimentation conducted under conditions never encountered in nature. Already in 1956, Payet, Camain, and Pène had rejected the role of food as a cause of PLC in FWA (Barret 1975). How could alimentary insufficiency be so discriminatory in its consequences? For while certain people succumbed to nutritional fatality, others in the same family, even in the same villages, experienced the same living conditions but never presented any sign of hepatic illness. These specialists showed that nutritional steatosis never led to cirrhosis and that people developing kwashiorkor had never developed cirrhosis.

As early as 1956, specialists from Dakar had proposed that viral infections were at the origin of PLC, without succeeding in proving their hypothesis. Baruch Samuel Blumberg's 1964 discovery of *Australia antigen* (a substance that causes the body's immune system to react by producing antibodies) confirmed their observations. So, after investigations in twenty countries, specialists could finally superpose the maps of PLC and *Australia antigen*. The discovery of this sub-group corresponded to a real epidemiological marker to identify all the patients suffering from hepatitis B, particularly in Africa and in Asia.

Conclusion

For decades, cancer had claimed many African lives without anyone being aware of its existence. For a long time it remained a low priority among colonial concerns because of a long-standing discourse organized around its absence among 'uncivilized peoples'. Although physicians increasingly accounted for cancer's distribution, instead of researching its real causes they lost themselves in its few known symptoms. For some physicians, like Le Dantec, Dupont, Delbet and Hoffman, cancer was only a disease of Europeans, or of industrially developed countries. Thanks to investigations conducted by other physicians – such as Montpellier, Gérard, Mouchet and Denoix – colonial medicine or biomedicine turned to the question of causes, including the factors contributing to PCL. The acknowledgment of many cancer cases among 'Black Africans' gradually dismantled the myth that cancer was a disease of Whites.

In the beginning, researchers focused their attention on the clinical state of sick people, as well as on the environment and on the factors that caused them to believe that PCL was limited to tropical regions. The discovery of the turkey X disease in the 1960s rapidly led to the belief that the cause of the disease was identified, without considering possible discrepancies between laboratory experiments and field tests. Biomedical specialists appeared to be lost, because the studies were increasingly complex and expensive but yielded no real results. Africa was abandoned to the ill effects of its environment, despite some success in the identification of alpha-feto proteins (AFP) (a major plasma protein contained in the human foetus) (Masseyef 1972) and the discovery of Australia antigen.

With the discovery of PLC among 'Black Africans', it was no longer possible to claim that 'Black indigenous people' in Africa did not suffer from cancer. The research that Professor Pierre Denoix (1947), in particular, conducted in French Africa from the 1940s proved that 'Black Africans', like other people in the world, were victims of cancers because cancer is a cell disease. The vaccine against hepatitis B created the conditions for the prevention of PLC.

Jean-Paul Bado (History PhD, Aix en Provence University, 'Health and Development' certificate, Marseille Medical School) is Director of the Group for Reflection and Action for Health in Africa (GRASA), and Senior Researcher at the University Aix-Marseille (IMAF). He is a member and reviewer of various medical publications: *Médecine Tropicale*, *Bulletin de la Société de Pathologie exotique*, etc. He specializes in the history of health, disease and medicine in Francophone Africa of the colonial and postcolonial periods. His publications include *Médecine coloniale et grandes endémies en Afrique* (1996), *Conquête de la médecine moderne en Afrique* (2006), and *Le docteur Eugène Jamot, le médecin de la maladie du sommeil* (1879-1937) (2011). Jean-Paul Bado's current work addresses the history of cancer and the history of concepts of medicines and disease in Africa. The title of his forthcoming book is: *Histoire du cancer en Afrique. Mensonges et véritiés.*

Acknowledgement

I wish to thank Rebecca Wilkin for improving the readability of my English.

Notes

1. Words used at the time.
2. French original: 'Les populations que nous avons visitées présentent un état de civilisation fort peu avancée et nos observations confirment l'opinion émise par certains auteurs, à savoir que le cancer est une maladie de civilisation (ou tout au moins de ce que l'on appelle la civilisation.) Ce qui est certain, c'est que çà n'est pas une question de race car

le cancer existe chez les Noirs d'Amérique, il existe également à Dakar où il serait aussi fréquent qu'en Europe. Or ces noirs d'Amérique comme ceux de Dakar ont la même nourriture que les blancs qui les entourent.'

3. Considered to be a vitamin of the B complex, it is the basic constituent of lecithin and prevents the deposition of fat in the liver.

Bibliography

Adam. 1924. 'Cancer de l'ovaire', *Bulletin de la société de pathologie exotique* 17: 330–331.

Adler, S. and E.H. Cumming. 1923. 'Malignant Growths in Natives of Sierra Leone', *Annals of Tropical Medicine and Parasitology* 17: 535–537.

Bablet, J. 1935. 'Le cancer dans les pays tropicaux', *Revue coloniale de médecine et de chirurgie* 70: 144–150.

——— 1947. 'Le cancer primitif du foie sous les tropiques', *Revue coloniale de médecine et de chirurgie* 138: 174–182.

Bado, J.P. 1993. 'De l'observation du sommeil à la découverte du trypanosome et de la glossine', *Médecine tropicale* 53(2): 149–157.

——— 1996. *Médecine coloniale et Grandes Endémies en Afrique 1900-1960: Lèpre, trypanosomiase humaine et onchocercose*. Paris: Editions Karthala.

——— 1999. 'Histoire, maladies et médecine en Afrique occidentale XIXème–XXème siècles', *Revue Française d'Histoire d'Outre Mer* (RFHOM) 322–323, pp. 237–268. Retrieved from http://www.persee.fr/web/revues/home/prescript/article/outre_0300-9513_1999_num_86_322_3727

Bado, J.P. and M. Michel. 1991. 'Sur les traces du docteur Emile Marchoux, pionnier de l'Institut Pasteur en Afrique noire', in M. Morange (ed.), *l'Institut Pasteur: contributions à son histoire, Histoire des sciences*. Paris: La Découverte, pp. 296–311.

Barret, C. 1975. *Les cancers primitifs du foie: 40e Congrès français de médecine Dakar*. Paris: Masson.

Bergeret, C. 1948. 'Un syndrome de dénutrition maligne de l'enfant noir "le kwashiorkor"', *Bulletin médical de l'AOF* 5: 257–261.

——— 1950. 'Quelques réflexions sur la cirrhose des Noirs', *Médecine tropicale* 2: 379.

Berman, C. 1941. 'The Pathology of Primary Carcinoma of Liver in the Bantu Races of South Africa', *South African Journal of Medicine and Sciences* 6: 11–26.

——— 1951. 'Carcinomes primitifs du foie chez les races bantou d'Afrique du Sud', *SHP* 2(26): 1034–1042.

Calmet, L. and Journe. 1946. 'Le cancer primitif du foie chez les tirailleurs sénégalais. Données anatomo-clinique', *Marseille médical* 83(3): 126–130.

Cazanove, F. 1931. 'Le cancer des Indigènes dans les colonies françaises', *Bulletin de l'Office international d'hygiène publique* 23(2): 880–891.

Commes, C. and H. de Vallende. 1916. 'Myxo-sarcome du bras chez un Bambara', *Bulletin de la Société de Pathologie exotique* 9: 702–703.

D'Anella, M. and R. Tourenc. 1938. 'La chirurgie aux colonies, possibilités chirurgicales du sud Cameroun (Yaoundé)', *Presse médicale* 32: 619–620.

Darmon, P. 1993. *Les cellules folles. L'homme face au cancer, de l'Antiquité à nos jours*. Paris: Plon.

Darnis, F. 1980. 'Le rôle des facteurs alimentaires dans la genèse du CPF', *Gazette médicale de France* 87(22): 2863–2870.

Darwin, C. 1859. *On the Origin of Species by Means of Natural Selection or The Preservation of Favoured Races in the Struggle for Life*. London: John Murray.

De Lavison, R. 1869. 'Chronologie des maladies de la ville de Saint-Pierre (Martinique) de l'année 1837 à l'année 1856', *Archives de médecine navale* 12: 335–364.

Delbet, P. 1936. 'Le cancer dans la race noire', *Bulletin de l'Académie de médecine* 12: 483–492.

Denoix, P. 1947. 'Le cancer en AOF de 1940 à 1946 d'après les documents recueillis par le service de santé des troupes coloniales', *Bulletin de l'association française pour l'étude du cancer* 1(34): 194–217.

Diop, B., F. Denis and P. Maupas. 1981. 'Epidémiologie du cancer primitif du foie au Sénégal', *Médicine d'Afrique Noire* 28: 217–223.

Diouf, M. 1967. *Répartition géographique du cancer primitif du foie en Afrique et ses relations avec la culture arachidière*, Ph.D. thesis. Bordeaux: Université de Bordeaux.

Dumas, J.M.R. 1939. 'Le cancer dans les colonies françaises', *Annales de médecine et de pharmacie coloniales* 37: 780–789.

Dupont, R. 1942. 'Le cancer au Tchad, en particulier chez les Saras', *Presse médicale* 51: 733.

Geyer, A. 1946. 'Aperçu sur la fréquence et les modalités du cancer en Afrique occidentale française, Réunion de médecine, de chirurgie et de pharmacie de Dakar', *Bulletin médical de l'AOF* 1: 13–14.

Guérin, J. 1952. *Contribution à l'étude du cancer primitif du foie chez l'Africain à Dakar*, Ph.D dissertation. Bordeaux 1951-1952 Nr. 153. Imprimerie Deniaud Frères.

Headrick, R. 1994. *Colonialism, Health and Illness in French Equatorial Africa (1885-1935).* Piscataway: African Studies Association.

Heckenroth, F. and E. Bergonier. 1923. 'Renseignements démographiques sur Dakar en 1922', *Bulletin de la Société de pathologie exotique* 16: 438–464.

Jambon. 1914. 'Variole et vaccine dans le cercle de Mono en 1914 (Dahomey)', *Annales d'hygiène et de médecine coloniales* 15: 828–838.

Jojot, C. and J. Laigret. 1922. 'Un cas de tumeurs superficielles multiples observées au Cameroun', *Bulletin de la Société de Pathologie exotique* 15: 956–958.

Jonchere, H. 1948. 'Contribution à l'étude du cancer en Afrique occidentale française', *Bulletin médical de l'AOF* 5: 247–256.

Kennaway, E.L. 1944. 'Cancer of the Liver in the Negro in Africa and in America (le cancer du foie chez les Noirs d'Afrique et d'Amérique)', *Cancer Research* 4(9): 571–577.

Kermorgant, A.1901. 'Quelques us et coutumes des indigènes de la Côte d'Ivoire', *Annales d'hygiène et de médecine coloniales* 14: 146–151.

Le Breton, E. 1958. 'Récentes acquisitions sur l'hépatome expérimental du rat', in F.C. Roulet (ed.), *Cancer primitif du foie et des voies biliaires.* Paris: Masson, pp. 62–98.

Le Breton E., J. Boy, E. Chany, C. Frayssinnet, A. Jacob, Y. Moule and A.M. De Recondo. 1963. 'L'hépatome expérimental du rat', in R. Dupuy (ed.), *Les tumeurs malignes du foie.* Paris: Masson, pp. 1–42.

Le Breton, E., C. Frayssinnet and J. Boy. 1962. 'Sur l'apparition d'hépatite "spontanée" chez le rat Wistar. Rôle de la toxine de l'Aspergillus flavus. Intérêt en pathologie humaine et cancérologie expérimentale', *Archives de l'Académie des Sciences* 255: 784–786.

Le Dantec, A. 1929. Précis de pathologie exotique, 5th edn. Paris: Gaston Doin & Cie.

Leitner, A.J. 1945. 'Le cancer en Côte française des Somalis', *Bulletin de la Société de pathologie exotique* 38(78): 235–241.

Manson, P. 1898. *Tropical Disease. A Manual of the Diseases of Warm Climates.* London: Cassel and Company.

Masseyef, R. 1972. 'Human alpha-feto-protein', *Pathologie Biologie* 20 (15–18): 703–725.

Monnier, Y. 1999. *L'Afrique dans l'imaginaire français, fin du 19ième – début du 20ème siècle.* Paris: l'Harmattan.

Montpellier, J.M. and J. Montpellier. 1947. *Le cancer en France d'outre-mer*. Algiers: Libraire Ferraris.

Mouchet, R. and P. Gérard. 1919. 'Contribution à l'étude des tumeurs chez les noirs d'Afrique centrale', *Bulletin de la Société de Pathologie exotique* 12: 567–581.

_____ 1926. 'Le cancer et les noirs de l'Afrique centrale', *Bulletin de la Société de Pathologie exotique* 19: 564–569.

Moustardier, G. 1937. 'A propos de deux cas de cancer du sein chez l'homme observé chez les indigènes à Madagascar', *Bulletin de la Société de Pathologie exotique* 30: 884–889.

_____ 1938. 'Sur la fréquence et les modalités du cancer à Madagascar', *Bulletin de l'Association française pour l'étude du cancer* 74(1): 24.

Ninard, B. 1950. *Tumeurs du foie*. Paris: Le François.

Ninard, B. and J. Hintermann. 1945. 'Les tumeurs de la travée hépatique chez le porc au Maroc', *Bulletin de l'Institut d'Hygiène du Maroc* 5: 49–57.

Nogue, M. 1920. 'Note sur la fréquence du cancer primitif du foie chez les indigènes du Sénégal', *Bulletin de la société de médecine et de chirurgie française de l'Afrique de l'Ouest africain* 1(6): 155–156.

Ortholan. 1909. 'Les Cancers dans les pays tropicaux', *Annales d'hygiène et de médecine coloniales* 12: 140–147.

Palès, L. 1946. 'Cancer et race', *Bulletin médical d'Afrique de l'Ouest* 3(1): 73–96.

_____ 1954. *L'alimentation en AOF, milieux, enquêtes techniques, rations, missions anthropologiques de l'Afrique occidentale françaises, avec la collaboration de Marie Tassin de Saint Pereuse, préface du professeur A. Mayer*. Dakar: O.R.A.N.A.

Payet, M., R. Camain and P. Pene. 1956. 'Le cancer du foie. Etude critique à 240 cas', *Revue internationale d'hépatologie* 6(1): 1–86.

Peers, F.G., G.A. Gilman and C.A. Linsell. 1976. 'Dietary Aflatoxins and Human Liver Cancer. A Study in Swaziland', *International Journal of Cancer* 17(2): 167–176.

Pene, P. and M. Boisson. 1975. 'Cirrhoses et cancers primitifs du foie', in C. Barret (ed.), *Les cancers primitifs du foie: 40e Congrès français de médecine Dakar*. Paris: Masson, pp. 77–88.

Peyrot. 1905. 'Us, coutumes, médecine des Bambara', *Annales d'hygiène et de médecine coloniales* 8: 456–473.

Phiquepal D'Arusmont, L. 1930. 'Les tumeurs malignes dans la race noire', *Bulletin de la Société de Pathologie exotique* 23(1): 109–114.

Pinell, P. 1992. *Naissance d'un fléau: histoire de la lutte contre le cancer en France, (1890-1940)*. Paris: Métaillé.

Ringenbach, J. and N. Guyomarch. 1915. 'Notes de géographie médicale de la section française de la mission de délimitation d'Afrique équatoriale française et Cameroun, 1912-1913', *Bulletin de la Société de Pathologie exotique* 8: 301–313.

Rongier, P.E. 1931. 'Tumeur primitif du foie avec cirrhose. Ascite hémorragique', *Lyon médicale* 148: 548–550.

Scott, D. 1965. *Epidemic Disease in Ghana, 1901-1960*. London: Oxford University Press.

Smith, E.C. and B.G.T. Elmes. 1934. 'Malignant Disease in Natives of Nigeria an Analysis of Five Hundred Tumours', *Annals of Tropical Medicine and Parasitology* 27(4): 461–512.

Spire. 1919–1920. 'La lèpre au Dahomey', *Annales d'hygiène et de médecine coloniales* 18-19: 166–172.

Thuillier, P. 1980. *Le petit savant illustré*. Paris: Seuil.

Tilley, H. 2011. *Africa as a Living Laboratory: Empire, Development, and the Problem of Scientific Knowledge, 1870-1950*. Chicago: University of Chicago Press.

Vaughan, M. 1991. *Curing Their Ills: Colonial Power and African Illness.* Cambridge: Polity Press.

WHO. 1980. *Mycotoxines. Critères d'hygiène et d'environnement 11.* Geneva: World Health Organisation.

6

Postponing Equality

From Colonial to International Nutritional Standards, 1932–1950

Maria Letícia Galluzzi Bizzo

From the 1930s onwards, international organizations started to monitor the world food situation and deliver recommendations about human nutrient needs. They defined standards of nutritional normalcy and, accordingly, of proper diet intake.

During the twentieth century, nutritional standards became a regulated way of speaking about food and hunger. They increasingly served as programmatic constructs in medical assessment, and were crucial to the planning of food supplies, rations and policies. The numerical quantification of nutrient requirements was useful for monitoring and controlling populations and for planning food policies; it contributed to constructing hunger as a social problem (Cullather 2007; Vernon 2007: 111). At the same time, nutrient recommendations were charged with social, ethical and historical meanings. This chapter examines how the episteme of human variation, in part originated from colonial practices, was articulated with social conjunctures, political contexts and tacit cultural assumptions in the shaping of an authoritative, international rule of difference regarding human calorie standards.

Attempts to cluster human groups became more complex in the nineteenth century, with four to six major races and several gradients (Tilley 2011). Nutritional science also linked 'primitiveness' to remote geographic regions, 'exotic' foods and habits, and 'backward' agriculture. Researchers perceived a great diversity of appearance, metabolism, genetics, evolutionary biology, adaptive physiology and habits in the twentieth century and so recommended more studies on race. Their intention was to understand whether races could exist or not, and to understand whether heredity or adaptability, or both, could play a part in the differences. This chapter shares with Barbara Cooper and Jean-Paul Bado (in this volume) a focus on core concepts in colonial medical perceptions of disease. Malnutrition, in particular, led to a concern with people's diets, ethnicities, poverty, environment and cultural 'backwardness', and thus involved the scrutiny of subjects' overall way of living.

First constructed in the 1930s by the Health Organisation of the League of Nations (HOLN), and later by the Food and Agriculture Organisation of the United Nations (FAO), calorie standards became instrumental in medical governance and science. Interestingly, scientists who had spent part of their career in the 'living laboratories' (Ruxin 1996) of the colonies proposed lower (in comparison with those delivered for Western people) and locally based (instead of international) calorie standards. And yet, they sought simpler methods to evaluate body nutritional status. The biological and social arguments that grounded it, however, were strikingly ambiguous.

Against this background, this chapter argues that nutritional science was heavily invested in the production of population difference, albeit in a way that does not sit easily with other difference regimes of the time. Nutritional science, dominated by international networks, was not grounded or ruled by racial science, and has always considered that environmental conditions could impact on human physiology. Nevertheless, outstanding scientists in charge of fixing international food recommendations proposed differential calorie allowances for non-Western people, which enabled a racialization of these groups. Among them were populations in severe need of more food. This racialization is not a racist ideology, but it still represents a deterministic way of seeing the 'other', in this case due to diet. The 'rice-eaters' category, frequently used by these scientists, defined people who had rice as the basis of their daily diets as attaining very low calorie intakes. It was also used in a generalizing way as a synonym for 'primitive' and 'tropical'. By assuming that these people shared a common biological feature, namely, being alive and apparently healthy despite their defective calorie intake, a new biological category was created, and although it did not directly refer to common racial representations, it marked a profound distinction between populations. This difference was sustained in supposed biological variations, particularly in the assumption of diverse metabolisms. At the same time, obvious commonalities – most significantly the basal metabolic rate – were hushed up, despite the already accepted scientific demonstration that calorie metabolism was basically the same in all humans. More importantly, in official documents the differential requirements were naturalized in their biological, social and political facets, and colonial rule was not seen as a possible contributor to poor diets. Whilst up to the late 1940s the leading actors forged this difference as inherent in the bodies of the people, new political and scientific trends paved the way for its revision. In what follows, I will develop this argument in greater depth, covering the period from 1932, when the first nutritional studies of the HOLN were published, to 1950, when such new trends commenced.

Food and Science

In recent years, historians have explored how feeding nations and territories became a crucial political, social and economic affair during the twentieth century. State-building, national food resources, military interests and labour productivity stimulated the development of calorie standards as numerical indices. Both Cullather (2007: 338) and Barona (2008: 89) indicate how the construction of nation-states and the role of the state as a social regulator made nutritional science part of the governance for national 'vitality' and a political issue. Dixon underlines the fact that this was extended to the colonial context (2009: 333). Friedmann and McMichael emphasize the role of financial interests in the political economies of national food systems (1989: 96). For the British Empire, Cullather shows that calories were employed to estimate the food resources at the disposal of the state (2007: 360) and Vernon demonstrates that nutritional knowledge helped to frame the political economy (2007: 97). Calories, as Cullather argues, contributed in the transformation of diet into an international issue (2007: 363).

Science was embedded in and actively contributed to these political economies. Colonial studies in the 1920s to 1930s were in part grounded in industrialization and labour efficiency interests (Lewis 2000: 132), but besides revealing disastrous nutritional conditions (Orr and Gilks 1931; Economic Advisory Council 1939), they provoked concerns about health in the colonies, and epistemologically tied together agriculture, economics and public health (Worboys 1988: 213; Hardy 1995: 66; Wylie 1999: 283; Bivins 2007: 547). As for the international sphere, a social and economic framework was constructed for food problems, with malnutrition becoming a politicized theme in the HOLN, as underlined by Borowy (2010).

Great Britain, with its active scene of science and policies of nutrition, played a crucial role in international nutritional thought. Since the nineteenth century, epidemics, war, manpower, economics and social medicine had stimulated studies on human growth and nutrition in the UK (Tanner 1981: 142; Vernon 2007: 56). From the 1920s, British scientists extended their investigations in the colonies to inform colonial policy (Wylie 1999: 281). According to Bivins, this work inspired approaches to fight hunger (2012: 3). The science of nutrition furnished governments with pragmatic mechanisms for governing hunger, a health problem that generated continuing stewardship in the colonies (Vernon 2007: 81).

War played a double role in creating a need for nutritional policies and studies. As indicated by Ebbs (Ebbs 1991: 79), during the Second World War calorie standards were indispensable in planning military diets and in bringing relief to liberated populations. Staples highlights the fact that such standards informed strategic bi- and multilateral food relief programmes in the Cold War, as well as informing the United Nations' monitoring of the world food production and consumption for humanitarian and trade reasons (2006: 137). Other studies show that after the Second World War, new views about poverty arose, particularly a more sympathetic rather than moral analysis of poverty (Lipton and Ravallion 1997: 64). For the same period, historians observe less linking of poverty and disease to 'primitiveness' (Cooper 1996: 120), increased international critiques of colonialism (Patil 2008; Schuknecht 2010: 58) and the invention of the 'development of the Third World' as an international paradigm (Escobar 1988: 440). Nutritional knowledge was permeated by these cognitive matrixes that decisively influenced the nutritional work in international organizations.

In the history of investigating human differences, biological traces were not the only criteria used to classify human populations: culture, geography, language, behaviour, economic conditions, techniques, food habits, diseases and biological adaptation to environment were likewise employed. One category used by nutritional scientists, but also by other medical scientists, to characterize Asian populations in the first half of the twentieth century was that of 'rice-eaters', that is, 'natives' whose diets yielded calories almost exclusively from rice. Arnold explores how the 'rice problem', as articulated by colonial investigators, was at the core of viewing malnutrition in India as a sign of diets that were faulty in calories and vitamins (1994: 12; see also Cullather 2007: 364).

This points to a long-standing cultural history of European prejudices against rice-based diets. As Arnold highlights, rice-diets had already long before 1900 been depreciated in European circles (1994: 12; see also Cullather 2007: 359). This can be substantiated by numerous citations from primary sources: 'Thousands of the beef-fed British govern and control ninety millions of the rice-eating natives of India' (Hale 1841: 31), stated one author as far back as 1841. 'As long as the Javanese live mainly on rice', another one claimed in 1852, 'they will be subjugated by the Dutch' (Moleschott 1852: 28).

The widespread idea that rice-diets were quantitatively and qualitatively defective was then only reinforced by experimental data, yielded in studies soon after the turn of the century (Mellanby 1919: 412; McCarrison 1924: 416), in particular studies undertaken after 1913 by the Northern Ireland physician Robert McCarrison (1878–1960) in Coonoor, India, and from 1918 by the Dutch physician Christiaan Eijkman (1858–1930) and the Dutch chemist Petrus Jansen (1884–1962) in Java, Batavia (Jansen and Donath 1926). Their respective successors, Wallace Aykroyd and André Gerard van Veen, are the main actors in this chapter. Besides being deeply involved with the 'rice-problem' in the colonies, they later triggered and influenced the construction of international differential calorie norms.

This chapter aims to describe and analyse the process of constructing international, differential calorie standards by colonial scientists, as linked to the 'rice-problem', and those scientists' endeavours to prescribe international 'soft-laws' regarding the culture of nutritional governance. It attempts to explore how numerical calorie indices were more a result of political choices, and thus historically contingent, rather than of biological evidence; the result, however, was a racialization of human groups. This chapter also hopes to give additional evidence of the continuities between the HOLN and the FAO in this terrain, to suggest some links between international and colonial nutritional thought, and to bring to light some details on the predominance of British nutritionists in this process.

Calorie Science up to 1932

Nutritional science in the 1930s was mainly characterized by studies directly or indirectly attempting to shed light on human requirements – mostly in the form of bioassays, that is, balance studies on anabolism, metabolism and catabolism –

on composition of foods, lactation and child growth, and on basal metabolism in different stages and conditions of life.[1]

The bases for calorie requirements had been construed since the eighteenth century, with the French chemist Antoine Lavoisier (1743–1794)'s oxygen combustion studies, and gained impetus in the nineteenth and early twentieth centuries through a transnational network of investigators: most prominently, the German scientists, the chemist Justus von Liebig (1803–1873), the physiologists Carl von Voit (1831–1908) and Max Rubner (1854–1932); the US-American chemist Wilbur Atwater (1844–1907); and the British physiologist Graham Lusk (1866–1932) (Hargrove 2006: 2959; Leitch 1942: 515; Beaton 1983: 340; Nichols 1994: 1719). British scientists delivered the first dietary standards during the 1862 economic depression to help the government establish the lowest purchase cost in order for the unemployed not to starve (Leitch 1942: 521). Estimates (including caloric expenses while working), posited since the nineteenth century by authoritative individual European and North-American investigators, were around 3,000 calories a day (FAO 1950). Those scientists were partially or completely funded by governments and advocated the incorporation of calorie requirements into policies, and advised governments accordingly (Hargrove 2006: 2961; Beaton 1983: 340; Nichols 1994: 1727). In the First World War, an Inter-Allied Food Commission estimated calorie requirements in order to ship foods to Europe (Bane and Lutz 1943). During the depression of the 1930s, dietary standards guided welfare allowances in several countries (Hargrove 2006: 2960; Barona 2008: 100). In the next global war, the accuracy of the numbers was important, as exemplified by the nutritionist in charge of the US military diets: 'The supply and distribution of food under government control in war time makes great demands on nutritional science. The calculation of caloric requirements […] assumes a new importance. […] Even a very small percentage of millions of tons represents an impressive amount of food so there is an understandable desire to refine caloric calculations to the utmost' (Keys 1945: 81).

In contrast to these seemingly universal claims, and depending on the population chosen as representative of 'normal' and thus referential, difference was nevertheless entrenched in calorie allowances, as I will demonstrate later on. Certainly, it was no minor problem for those who were to fix these standards for their application in population policies: if inadequate calorie standards were used to plan diets, body status impairment and defective health could occur. To establish how many calories were needed, several factors were taken into account: organic expenditure to maintain basic physiological functions and to perform intentional muscular work (the basal metabolic rate), sex, age and body size. By the early 1930s there were widespread indications that the innate basal metabolism did not significantly differ between inhabitants of the tropics and of temperate regions, and that climatic rather than racial influences explained the slight differences found (Eijkman 1896; Almeida 1920; Steggerda and Benedict 1928; Mason and Benedict 1931).

Today, hunger and impaired body status are viewed as historical indicators of unequal societies (Bogin and Keep 1999: 343; Destombes 2006: 200; Vernon 2007: 39; Steckel 2008: 19). But already back then, the role of living conditions for human body variation had long been recognized (Tanner 1981: 102; Floud, Gregory and Wachter 1990: 86). Studies on colonial nutrition had detected how faulty diets impaired body weight and/or height; in some African colonies exhausting labour conditions had been a suspected major cause of malnutrition (Rivers 1996; Wylie 1999: 60; Vernon 2007: 77). Classical theories from 1919 – the Harris-Benedict calculation on the basis of height and weight (Harris and Benedict 1919) – and even from the 1880s – incorporating body surface area (Rubner 1883; Richet 1885) – had uncovered the crucial role of body size and surface for the basal metabolism. Furthermore, the effects of low dietetic patterns (Benedict and Cathcart 1913) and hard labour (Keys 1930) on the metabolic rate were also well recognized. Therefore, historical conditions of life were acknowledged by science as key influences on human calorie expenditure. And yet, differential standards seemed indispensable to those who came to be crucial actors in this field of interactions between science and policy makers.

Steps towards Differential Standards

The early 1930s saw intensified attempts by international organizations to monitor the world food situation, and to deliver recommendations about human nutrient needs. Scientists took on leading roles in defining standards of nutritional normalcy and, accordingly, of proper diet intake. Particularly in two important international organizations, the HOLN (1932–1939)[2] and the FAO (from 1945 onwards), such work was led by colonial scientists, who had spent a significant part of their careers as doctors or medical scientists in the colonies of European empires.

Due to the world economic recession initiated in 1929, governments became increasingly worried about international – mainly European – health and labour productivity. Accordingly, in the early 1930s the HOLN saw a need to found and disseminate scientific and political thought that would bind nutrition to public health (Barona 2008: 99). In 1931, a British physician with colonial experience, Wallace Aykroyd (1899–1979), was assigned to the task; his work would soon be recognized worldwide. In the colonial context, Aykroyd had conducted studies regarding the 'rice-eaters'. In 1936, he would invite the Dutch colonial biochemist André Gerard van Veen (1903–1986) to help him in the HOLN. Van Veen was then secretary of the Indonesian Science Council and a specialist on the 'rice problem'.

As members of a medical colonial elite in the 1930s, both Aykroyd and van Veen had embarked on laboratory experiments, respectively in India and Batavia, and continued research traditions initiated by their renowned predecessors on the links between rice-diets and beriberi (the deficiency of vitamin B1). In 1935, Aykroyd replaced Robert McCarrison at the Nutritional Research Laboratories in the city of Coonoor, and van Veen succeeded Barend Jansen at the Eijkman Institute of Nutrition, Java. Both studied metabolism, chemical composition of foods, isolation of nutrient substances, dietary consumption and somatometric conditions.

As a leader in one of the most powerful international organizations, Aykroyd was responsible for inviting skilled scientists to participate in advisory expert groups, whose opinions were negotiated and whose recommendations were submitted to the Assembly. Once approved, their recommendations acquired the status of 'soft-laws', non-coercive cultural models of governance and scientific procedure (McNeely 1995: 127; Barnett and Finnemore 1999). The advisors were not chosen by member-nations, but by the international organizations' scientific leaders. This social contact propitiated pre-delimitation of shared objects and meanings, and played a remarkable role in the organizations' statements. As a result, the colonial powers predominant in the HOLN nurtured an active process of inter-imperial borrowing, and maintained a high interest in colonial medicine (Borowy 2010).

Aykroyd was part of the HOLN's staff from 1931 to 1935, later remaining highly influential as an advisor. In 1935, at the request of the HOLN, he co-authored the classic paper *Nutrition and Public Health*, attributing great complexity to human nutrient needs and beginning to touch on the theme of colonial undernourishment (Burnet and Aykroyd 1935). He was ambiguous about biological difference: he affirmed that race had little influence over the dietary organic utilization, but suggested that investigations on different races were needed urgently. He also maintained that parameters of 'normal' body weight should be based on healthy, biologically well-developed individuals (Burnet and Aykroyd 1935: 238).

As nutrition assumed an outstanding position in the HOLN's agenda, in 1936 a Technical Commission on Nutrition was set up. Aykroyd successfully manoeuvred the 'rice-problem' – especially the 'Eastern' one – into place as a paramount theme. Some of the chief assertions about nutrition problems in HOLN official documents of the period matched or even reproduced Aykroyd's and van Veen's writings (Technical Commission on Nutrition, Sub-Committee on Nutrition in Asia and Tropical Countries 1938a). In 1936 the agency's Assembly requested studies on Asian and tropical diets (Technical Commission on Nutrition 1937a), a study mission to the Far East and the creation of the Sub-Committee on Nutrition in Asia and Tropical Countries. Aykroyd led this, with the help of van Veen. Two seminal publications separately authored by them in the early 1930s – *The Poor Rice Eater's Diet*, by Aykroyd, and *The Rice Problem*, by van Veen – were reprinted (unfortunately with no mention to their original reference data) and provided fuel for the Sub-Committee's debates and paved the way for the fixation of differential standards for colonial peoples. This occurred at the HOLN's Intergovernmental Conference of Far-Eastern Countries on Rural Hygiene in 1937 (Health Organisation of the League of Nations 1937). Held in van Veen's site of colonial research – Bandoeng, Java – the conference's agenda was decisively moulded by Aykroyd and van Veen and prioritized the 'rice-problem' (ibid.). Based on Aykroyd's and van Veen's recommendations of lower calorie standards for the 'natives', the conference entered history by setting the first international differential nutrient standards for non-Western peoples. This knowledge looked like it was international, but was not: very specific scientists, in specific positions, constructed it, through their negotiation of the local and the international.

Both Aykroyd and van Veen assigned poverty and education (the foremost explanations in the HOLN for malnutrition) and local conditions as causes and solutions for the 'rice-problem'. As for biological difference, Aykroyd had recurrent but mild doubts on the existence of a 'racial physiology' and its nutritional effects. Van Veen hypothesized that Java's people were functional despite their insufficient diet, maybe due to physiological adaptation (Technical Commission on Nutrition 1939b). Accordingly, he proposed a 'habit-norm' standard for non-Westerners, based on actual consumption (ibid.: 2). Van Veen was the only one in the HOLN to follow this biologically-reductionist line to justify minimum dietary standards for the Far East. Although Aykroyd conjectured on biological differences, he placed greatest emphasis on 'practical possibilities' of raising the dietary intake, which was in line with the British scientific tradition. He also considered that attempts to change dietary habits only through education would be 'insulting' (1933: 150). Van Veen, in line with trends in Dutch colonial medicine (Soekirmann et al. 1992), believed that '[in the Far East] the minimum requirements should be taken [...] with a view to attaining higher standards later on, as the population develops mentally and economically and nutritional enlightenment spreads' (Technical Commission on Nutrition 1939c: 2). The British influence in the HOLN's nutrition campaign made Aykroyd's positions predominate.

HOLN Standards

In 1932, at the HOLN's request, Aykroyd produced a ground-breaking paper connecting economy and health which brought calories to the fore (Aykroyd 1932). In 1935 HOLN experts, including Aykroyd, delivered the first international calorie table, based on Western men: 2,400 a day plus 75-300 per hour of (light, moderate or hard) work (Technical Commission of the Health Committee of the League of Nations 1936). On average it totalled 3,000-3,200 calories as a daily intake for men, a value re-endorsed by the HOLN in 1936 (League of Nations 1936).

Paradoxically, although the previous HOLN scales had been declared 'universally-applicable', in 1937 the HOLN's Bandoeng Conference of Far-Eastern Countries on Rural Hygiene, convened by Van Veen, recommended 2,600 calories for non-Western subjects, based on the 'Scale Adopted in the Southern Parts of India', originally developed by Aykroyd and his team at the Nutrition Laboratories in Coonor (Health Organisation of the League of Nations 1937: 105). The calorie recommendation in the index was superior to the disturbingly prevailing pattern of actual calorie intake (800–1,600) but inferior to that set up for the Occident. The difference was officially reasoned on:

- 'Practical conditions': the intake was so low, and the possibilities of amelioration so scarce, that a slow rhythm of improvement should be expected;
- Lack of physiological studies, especially about basal metabolism;
- Kind of labour: agricultural and manual occupations were defined as light work.

This inaugurated the HOLN's and later the FAO's practices of taking research data from Madras (nowadays Chennai, in India) and Java (in Indonesia, then Batavia) to generalize positions for vast parts of the world, by considering 'rice-eaters', 'tropical peoples' and 'natives' as synonyms. The 'rice-eaters' category was malleably used to strategically refer to a minority (the 'non-Western') and to a majority (the 'Tropics', or 'half of humankind'), exposing a scientific management of arguments operational in producing a particular new order, and a reflex of the relatively autonomous international organizations' self-referential knowledge.

Although the term 'race' appeared in the HOLN publications on nutrition in a sparse fashion and with irregular meanings, the term 'rice-eaters' and its synonyms exposed a flexible race-type categorization. Most notably, nobody in the HOLN, including Aykroyd and van Veen, scrutinized the political role of colonial rule in the impairment of feeding, body size and physiology in the colonies. Interestingly, that progress towards improved nutrition could only be slow, and that local conditions should solve the problem, was also stated in British documents (Colonial Office 1936). In a note to the HOLN, the Department of Public Health of the Government of British India indicated that the aims of investigations of nutritional problems in India differed from those in Europe or the United States because the diet fell so far short of the ideal that minimum, instead of ideal and unattainable standards, should be used (Technical Commission on Nutrition 1937b).

The differential 2,600 calories set up in 1937 would become a permanent one in the HOLN, being reaffirmed in 1938 and 1939 (and afterwards in 1946 and 1949 by the FAO). A 1938 report acknowledged the starvation diets, but postponed to the future the task of making them 'less remote' from the optimum (Health Section of the League of Nations 1938: 2). From the biological point of view, however, the explanations offered in the HOLN were erratic, sometimes pointing to a lack of studies, sometimes pointing in the opposite direction, as stated in a 1938 meeting of the Sub-Committee on Nutrition in Asia and Tropical Countries: 'The food requirements of the peoples of the Orient and Tropical countries do not essentially differ from that of the Occident civilization. Comparative studies on physiological differences due to race and climate […] are undoubtedly adequate […] However, it seems not probable that such studies reveal […] differences of racial physiology' (Technical Commission on Nutrition, Sub-Committee on Nutrition in Asia and Tropical Countries 1938b).

The next step in the HOLN's agenda on nutrition was an attempt to better advise national nutrition authorities on field studies. In 1938 – assigned after Aykroyd's suggestion – Édouard Bigwood (1891–1975), an outstanding Belgian physiologist on the nutrition of the Belgian Congo and professor of biological and food chemistry in the Faculty of Medicine of Brussels (Bigwood and Trolli 1937), was commissioned to write a methodological guide (Bigwood 1939). Yet for 'tropical peoples', it was agreed that a separate guide should be written (a task never accomplished). Bigwood's work became an internationally recognized reference. It was not uncommon for specialists trained in the colonies to advocate adapted methods for the tropics (Bivins 2012: 19); Bigwood followed this trend and stated that fewer and simpler methods to assess the state of nutrition should be used in such areas (1939: 49). With respect to calories, Bigwood simply repeated the Bandoeng recommendations (ibid.: 105), alleging that the size and growth of newborn non-Western babies was smaller (ibid.) and that in Batavia a man weighed 50kg instead of 65-70kg (ibid.: 97). This converged with van Veen's assertion that the Batavian 'natives' were 'relatively short in stature, the adults as a rule weighing 42-49kg' (Technical Commission on Nutrition 1939b: 2). It is interesting to compare this with a comment of Aykroyd's at that time: 'If the diets [...] are worked out in terms of protein, minerals, and vitamins, and the results compared with the standards suggested by the Technical Commission on Nutrition and other standards drawn up by physiologists, [...] the rice-eater's diet falls short of such standards in almost every important constituent' (Technical Commission on Nutrition 1939a: 3).

In 1939 the Technical Committee on Nutrition declared that rice-diets were the food of half of humankind; accordingly, in 1940, a massive work on nutrition in the Tropics and the Far East was scheduled, and the HOLN's journal was meaningfully dedicated to rice, republishing Aykroyd's and van Veen's works on this topic.[3] Copies of Aykroyd's paper were requested by the British Colonial Office to be circulated to colonial governments (Gautier 1941). In the meantime, however, the nutritional agenda was aborted as the Second World War broke out and the League declined politically, and was closed just after the war's end.

FAO Standards

The HOLN's and US-American endeavours in 1943–1945 were central to the creation of the FAO in 1945. War-time food shortages led the agency to resume the HOLN's nutritional work (Aykroyd 1947c). Invited by the director-general, Aykroyd was installed as the first director of the FAO's Division of Nutrition, with van Veen as his second man. The 'rice' issue was immediately put back on the agenda and soon declared a priority issue in the FAO (FAO Standing Advisory Committee on Nutrition 1946).

An ad hoc Committee on Nutrition was set up and, as a member, Aykroyd suggested that higher nutritional standards – such as those put forward by the HOLN or by North-American committees – should be applied to the United States and certain other countries. In tropical areas, they should stand as an ultimate objective. He proposed that a schedule drawn up in 1944 under his leadership of the Indian Research Fund Association should serve as a guide; it consisted of a set of cheap Indian foodstuffs and should be adapted to other tropical places using local foods as substitutes. He additionally suggested, reiterating the HOLN's positions, that a minimum of 2,500–2,600 calories should be consumed (1946a: 9). In practice, this repeated the HOLN's standards previously fixed on the basis of Aykroyd's and van Veen's suggestions. His proposal was criticized in India: 'The Indians attacked us (on political grounds) for selecting too low a figure, while those who have actually studied the nutrition problems of the [...] tropical races tended to hold the opinion that 2,550-2,650 per caput was too high. The level set was in my opinion a reasonable one, in relation to present intake and emergency subsistence levels' (1946b).

Consequently, some months later the FAO's standards for national emergency subsistence diets (to be used in natural disasters, for example) were settled by Aykroyd at no less than 2,000 calories for Europe and 1,600 calories for the Orient (1946c).

In 1947 a committee of experts on nutritional problems in the tropics and Asia was established (Aykroyd 1947e). As members, Aykroyd and van Veen sent a joint letter to the nutrition centres of the Far East, asking for data and publications (Aykroyd 1947d). Since several colonial territories were involved, Aykroyd also had extensive consultations with the British physiologist Benjamin Platt (1903–1969), nutrition advisor to the British Colonial Office, and vastly experienced on 'rice matters' (Aykroyd 1947a). Platt, assigned on Aykroyd's recommendation, was an outstanding member of the FAO Standing Advisory Committee on Nutrition and the FAO Nutrition Committee for the Far East; in the 1930s he advised the HOLN (Aykroyd 1948d). Platt and Aykroyd shared a past on British nutrition. The resulting publication was the first of the FAO Nutritional Studies series: *Rice and Rice Diets – A Nutritional Survey* (1947), which Aykroyd considered owed much to his and van Veen's colonial experience (Aykroyd 1948b). The work prompted the establishment of an FAO Regional Office for Asia and the Far East, and a Survey Mission to the Far East – with Platt as an 'essential member' (Aykroyd 1947b); the mission concluded that malnutrition was the usual condition among local children (Parran and Lakshmanan 1948).

In 1949, the FAO convened its first Expert Committee on Calorie Requirements. Aykroyd invited Bigwood (Aykroyd 1948c) and van Veen, but the latter was unable to take part.[4] Calorie needs were now fixed at 3,200 for individuals comparable to the reference point of a 65-70kg man (FAO 1950: 11). This, however, evaded the question of whether a lower body weight was to be seen as a natural given. Whether someone with a significantly lower body weight might simply be undernourished, and whether then a relative calculation of calorie need per kilo would be utterly misleading, remained unconsidered. Moreover, for 'different' populations, the HOLN's 1937 Bandoeng recommendations were once again reaffirmed, allegedly owing to a lack of biological studies (FAO 1950: 6).

In the same year, a Joint FAO/WHO Committee on Nutrition was created, and reiterated the differential methods of nutritional evaluation proposed by the HOLN (Joint FAO/WHO Expert Committee on Nutrition 1949: 49). While British administrators claimed that 3,200 calories were too low for its nationals (Joint FAO/WHO Expert Committee on Nutrition 1954), a prominent scientist close to Aykroyd[5] criticized the maintenance of the policy of difference (Passmore 1950); Aykroyd replied to him:

> The Committee was aware of the danger that their recommendations might be used to justify undernutrition [...] We don't really know the part played by heredity in determining stature and weight, [but] I doubt whether there

> is full justification for your statement that the vast majority of the human race are or were 65kg-persons. The [FAO] example you quote – i.e. that in which the weights of adult males […] are taken to be 50 kg […] – is hypothetical. (1950b)

New Trends

Platt would play a significant role in the FAO policies (Hardy 1995: 62). In the late 1940s he investigated an acute deficiency of proteins in young children – *kwashiorkor* – and created a United Nations' demonstration area in the Gambia. By 1952 he was a world authority in kwashiorkor (Ruxin 1996: 38). His participation in the Calorie Committee and behind the scenes at the FAO contributed to the substitution of calories by proteins as the main nutrient focus of the agency (Waterlow 1955). The Joint FAO/WHO Committee on Nutrition no longer highlighted malnutrition as a chronic lack of calories, but as an acute shortcoming of proteins. In comparison with the 'Asian' period, the FAO's geographical focus was expanded: 'One of the most widespread nutritional disorders in Tropical and Sub-tropical areas is a syndrome […] known […] as *kwashiorkor*' (Joint FAO/WHO Expert Committee on Nutrition 1949: 2). Aykroyd did not think proteins were a major problem in the 'rice-eating areas' (1950a). However, subsequently, 'rice' assumed a secondary role in the agency, including operational funds (Aykroyd 1949a).

During the 1940s, although new trends were added to continued ones in the field of nutrition science – like lactation, growth and balance studies – what occurred as a novelty inside the FAO was more an official acceptance of already-established theories than new discoveries. In both the HOLN and the FAO a fully informed science had always been practiced. Thus changes in the FAO's trends resulted from political changes in the United Nations regarding races and human types, and of new scientific paradigms regarding malnutrition.

In the scientific field, Max Rubner's work on bioenergetics in 1902 had already demonstrated body temperature climatic regulation; but only now were the effects of climate over basal metabolism well recognized (Albagli 1939). Evidence that the factor of race did not importantly affect the body's basal caloric expenses had been found by van Veen's predecessor in the colonies and were only now acknowledged (Eijkman 1896). The influence of previous diet pattern on basal metabolism was also reaffirmed during that time (Treichler and Mitchell 1941). The novelty was, therefore, that the FAO policies now incorporated theories well established outside of its own scientific activities. Nevertheless, scientific studies on nutritional relationships created a space for a better comprehension of the interdependence between the energy and the protein metabolisms, and the action of proteins on growth (Barnes, Bates and Maack 1946).

There were also changes in the international scientific and political context towards race. The 1948 United Nations' Universal Declaration of Human Rights and the UNESCO statements of the early 1950s on race transcended racial difference through the notion of a 'universal man' (Haraway 1989; Selcer 2012) and even redeemed the value of the 'first man', the African hunter-gatherer (Haraway 1989: 54). The ideology of 'freedom from want' (part of the North-American Salvationist discourse of freeing the world from material and symbolic restraints) pervaded the United Nations' agenda (Staples 2006: 83), and the organization emphasized the watching of food availability in both nations and territories (ibid.: 181). The premise of overcoming the 'underdevelopment' of the 'Third World' reinforced poverty (in its social and moral spheres) rather than 'difference' as the cause of malnutrition; provision of adequate food was declared a human right (United Nations 1948). The United Nations' and their agencies' discourse moved towards an expansion of the recommendation of more equity and universalism in the access of all populations of the world (regardless their geographical, biological, or political condition) to adequate diets.

As a result of these political and scientific trends, the report of the 1949 FAO Committee on Calorie Requirements, although editing for the last time in history the HOLN's differential standards, contained changes that would open the possibility for more equity in future international nutrition standards. The report explicitly recognized that the basal metabolism of non-Western people showed little or no variation in comparison to those of Western men. It also terminated the role of race over energetic metabolism: 'race per se does not influence calorie requirements' (FAO 1950: 8). Acknowledging that body or constitutional type influenced the basal metabolism, the committee added body surface as a variable in the calculations. It recognized that the differential standards were based on observations of actual rather than ideal consumption, thus reflecting physical requirements 'plus practical possibilities' (ibid.: 5) and that the differential scales were biased.

Conclusions

This chapter has demonstrated how the episteme of human variation, partially accrued from colonial science, was articulated with social, political and cultural assumptions, in the construction of a legitimized, international rule of difference concerning human calorie standards. International numerical calorie indices were charged with historical and ethical meanings.

The positions adopted by international organizations cannot be interpreted in only a negative way. Field science inserts practitioners into existing social relationships (Widmer 2012: 79), and they evidenced genuine concern with inadequate feeding as a sanitary and humanitarian problem (Borowy 2010). Aykroyd affirmed that famine and food shortage in India were accentuations of the routine situation (1946a: 6) and that clinical conditions in prison camps matched that of normal times among 'rice-eating' populations (1948a). It means that hunger was the common, condition of the daily life of the inhabitants, and natural disasters like droughts just worsened what was already the daily food reality of those populations. These actors called attention to the food conditions in the colonies and helped to integrate the territories into the international health discourse (Borowy 2009: 236).

Nevertheless, the choice of lower instead of 'universal' standards was segregating. If those populations were insufficiently known, would it not be fairer to adopt the universal parameters? Because the justifications for the lower caloric standards were hazy and paradoxical, they acquired transparency: their frailty exposed the power behind them. Interestingly, in both the HOLN and the FAO the general message was that national states should incorporate nutrition science as best practices of governance. But this discourse had its inner rationality disrupted by the construction of a contrary rule for colonial peoples. The 'practical possibilities' assertion is a syntax of difference, a paradox between biological, social and political realms. The scientists' constrained agency contributed to the apparent imprecision of opinions: colonial nutrition did inform international nutrition. Internationalism was a fragmentary, non-democratic kind of communality.

The postponing of equality – tacit in the attainment of lower, 'intermediate' standards – gives to both the past and future history of such regions a different cadence from that of the West. It is impregnated with the idea of superiority of the Occident (Said 1978) and with a notion of scarcity of material and symbolic conditions to allow progress. In practice, the differential recommendations supported the adoption of lower nutritional parameters at least in the colonies linked to the main scientists involved in the construction of these standards at the HOLN and the FAO, such as British, Dutch, French and Belgian territories.

The view of the social determinants of health is dependent on the interpreter's political, intellectual and geographic locus (Randall 2009). Colonial rule was not considered as a possible reason for nutrition problems inside the HOLN and the FAO; local causes and endogenous solutions were the preferred route, in a singular contextualization that detached the feeding problems from their historical connections. This singular contextualization exposed the complex mixture of politics and science as legacies of a particular social and political scenery.

To ask a question with a certain grammar is to be already engaged in a way of seeing (Wittgenstein 1953: 308). The tropics appeared more as a social than as a geographical frame. The politics of naming – the 'rice-eaters' and its synonyms – relationally 'formed' the colonial world. Categories can also be flexible or ambiguous (Bauer and Wahlberg 2009: 5). Ambiguity is not uncommon in the history of difference in biological sciences (Whitmarsh 2008: 1; Lipphardt 2012: 70; Widmer 2012: 69). While often supressed in the historiography, contradictory motives exist and are an integral part of the medical thoughts and practices (Whitmarsh 2008: 2). Ignoring such contradictions may prevent us from seeing historical evidence of the links between science and politics.

Maria Leticia Galluzzi Bizzo is a Professor at the Graduate Program on the History of Sciences and Technologies & Epistemology, Federal University of Rio de Janeiro, Brazil. She has been a scholar in residence at the Smithsonian Institution (Washington, D. C., USA) and has received research funding from the Fulbright Commission. She will soon be a resident scholar at the University of Macerata (Italy).

She has also been a visiting scholar at the Max Planck Institute for the History of Science (Berlin, Germany) and at the Ruralia Institute, University of Helsinki (Finland). She holds a PhD in the History of Sciences and Health (Fiocruz, Rio de Janeiro/Università di Roma 'La Sapienza', Italy) and has been an intern at the United Nations's FAO Archives (Rome, Italy). Her recent publications include 'Latin America and the "International Geneva: during the Interwar Period' (2015) and 'Highlights in the History of Coffee Science Related to Health' (2015). Her research interests encompass global history. transnational scientific relationships, and history of the human rights.

Acknowledgements

This chapter is based on sources from the League of Nations Archives (LNA) and the FAO Archives (FAOA), to whose teams the author is indebted. The work benefited from discussions during her visiting scholarship at the Max Planck Institute for the History of Science's research group 'Twentieth Century Histories of Knowledge about Human Variation'. She thanks Drs. Veronika Lipphardt, Alexandra Widmer, Ricardo Ventura, Jenny Bangham, Samuël Coghe, Lara Keuck and Yuriditzi Montijo (MPIWG), for their invaluable suggestions, as well as Drs. Flavio Edler and Marcos Chor (Oswaldo Cruz Foundation, Rio de Janeiro), for their generous reading of the first version of this work, when it was intended to be just a conference presentation. This work is dedicated to Eduardo Galluzzi Bizzo.

Notes

1. These were the predominant issues identified in the articles of the *Journal of Nutrition*, a classic in the field.
2. Acting since 1924, the HOLN was a technical body of the League of Nations (1920–1946), in charge of disseminating recommendations and data about health. It exerted an outstanding global influence on scientific and governmental trends (Borowy 2009).
3. In the *Bulletin of the Health Organisation* IX(3), 1940–1941.
4. Van Veen was assigned to work in the Netherlands' reorganization of the training on nutrition, but would be back to the FAO by 1950, becoming responsible for issues related to the Far East, and later heading the Food Science and Technology Branch, from which he retired in 1962. Nevertheless, during this gap he kept in close touch with the FAO Nutrition Division, for instance with regard to the designing of the Baguio Conference (Aykroyd 1949b).
5. The British physician Reginald Passmore (1910–1999) was a renowned expert on energy metabolism. He published more than 115 scientific papers. In 1937, Passmore joined the Indian Medical Service as an assistant director of the Nutrition Laboratories in Coonor, then headed by Aykroyd, with whom he became close friends and developed several studies, including investigations on the 'rice problems'. In 1946 he was assigned to replace Aykroyd at Coonor. Passmore took part in the Expert Advisory Panel on Nutrition for the WHO from 1951 to 1956, after being recommended by Aykroyd. He went on to

teach physiology and social medicine at the University of Edinburgh until his retirement (Anonymous 1999).

Bibliography

Albagli, B. 1939. *O Metabolismo Básico em Função da Alimentação e do Clima.* Rio de Janeiro: Atheneu.

Almeida, M.O. 1920. 'Le Métabolism Basal de l'Homme Tropical', *Journal de Physiologie et de Pathologie General* 18(1): 958–964.

Anonymous. 1999. 'Obituary: Dr. Reginald Passmore, 1910–1999', *Journal of Human Nutrition and Dietetics* 12(6): 533–534.

Arnold, D. 1994. 'The Discovery of Malnutrition and Diet in Colonial India', *The Indian Economic and Social History Review* 31(1): 1–26.

Aykroyd, W. 1932. 'Economic Depression and Public Health', *Quarterly Bulletin of the Health Organisation of the League of Nations* 1(1): 425–476.

——— 1933. 'Diet in Relation to Small Incomes', Quarterly Bulletin of the Health Organization 2(1-2): 130–153.

——— 1946a. *Report to the Ad Hoc Committee on Nutrition: Note on Food Requirements for India. Ad Hoc Committee on Nutrition. Report 1. February 27, 1946*, Washington: FAO. FAOA.

——— 1946b. Memorandum to D. Lubbock. Washington, 24 October 1946. FAOA.

——— 1946c. Memorandum to A.M. Acock. Washington, 14 November 1946. FAOA.

——— 1947a. Memorandum to the director-general, John Boyd Orr. Washington, 3 January 1947. FAOA.

——— 1947b. Memorandum to F.L. McDougall. Washington, 13 October 1947. FAOA.

——— 1947c. Memorandum to F.L. McDougall. Washington, 22 October 1947. FAOA.

——— 1947d. Memorandum to van Veen. Washington, 13 November 1947. FAOA.

——— 1947e. Memorandum to T.V.N. Fortescue. Washington, 21 November 1947. FAOA.

——— 1948a. Letter to Reginald Passmore. Washington, 1 April 1948. FAOA.

——— 1948b. Letter to Reginald Passmore, 1 April 1948. FAOA.

——— 1948c. Letter to Bigwood. Washington, 2 April 1948. FAOA.

——— 1948d. Memorandum to F.L. Wormald. Washington, 21 January 1948. FAOA.

——— 1949a. Memorandum to the acting director-general. Washington, 23 March 1949. FAOA.

——— 1949b. Letter to P. Peissi. Washington, 21 September 1949. FAOA.

——— 1950a. Letter to H.C. Trowell. Rome, 1 November 1950. FAOA.

——— 1950b. Letter to Reginald Passmore. Washington, 10 November 1950. FAOA.

Bane, S.L. and R.H. Lutz. 1943. *Organization of American Relief in Europe 1918-1919, Including Negotiations Leading up to the Establishment of the Office of Director General of Relief at Paris by the Allied Powers: Documents.* London: Oxford University Press.

Barnes, R., M. Bates and J. Maack. 1946. 'The Growth and Maintenance Utilization of Dietary Protein: Five Figures', *Journal of Nutrition* 32(5): 535–548.

Barnett, M. and M. Finnemore. 1999. 'The Politics, Power, and Pathologies of International Organizations', *International Organizations* 53(4): 699–732.

Barona, J.L. 2008. 'Nutrition and Health: the International Context during the Inter-war Crisis', *Social History of Medicine* 21(1): 87–105.

Bauer, S. and A. Wahlberg. 2009. 'Introduction: Categories of Life', in S. Bauer and A. Wahlberg (eds), *Contested Categories: Life Sciences in Society*. Farnham/Burlington: Ashgate, pp. 1–13.

Beaton, G. 1983. 'Energy in Human Nutrition: Perspectives and Problems', *Nutrition Reviews* 41(11): 325–340.

Benedict, F.G. and E.P. Cathcart. 1913. *Muscular Work*. Washington: Carnegie Institution of Washington.

Bigwood, E. 1939. *Guiding Principles for Studies on the Nutrition of Populations*. Geneva: League of Nations.

Bigwood, E. and G. Trolli. 1937. *Problème de l'Alimentation au Congo Belge*. Brussels: Imprimerie des Travaux Publics.

Bivins, R. 2007. 'The English Disease or Asian Rickets? Medical Responses to Postcolonial Immigration', *Bulletin of the History of Medicine* 81(3): 533–568.

——— 2012. 'Coming Home to (Post)colonial Medicine: Treating Tropical Bodies in Post-war Britain', *Social History of Medicine* 26(1): 1–20.

Bogin, B. and R. Keep. 1999. 'Eight Thousand Years of Economic and Political History in Latin America Revealed by Anthropometry', *Annals of Human Biology* 26(4): 333–351.

Borowy, I. 2009. *Coming to Terms with World Health: the League of Nations Health Organization, 1921-1946*. Frankfurt: Peter Lang.

——— 2010. 'The League of Nations Health Organisation: from European to Global Health Concerns?'. Retrieved 18 October 2014 fromhttp://www.academia.edu/798208/The_League_of_Nations_Health_Organisation_from_European_to_Global_Health_Concerns

Burnet, E. and W. Aykroyd. 1935. 'Nutrition and Public Health', *Bulletin of the Health Organization of the League of Nations* 4(2): 232–474.

Colonial Office. 1936. *Nutrition Policy in the Colonial Empire. Dispatch from the Secretary of State for the Colonies, dated 18th April, 1936*. London: H.M.S.O.

Cooper, F. 1996. *Decolonization and African Society: The Labour Question in French and British Africa*. New York: Cambridge University Press.

Cullather, N. 2007. 'The Foreign Policy of the Calorie', *American Historical Review* 112(2): 337–364.

Destombes, J. 2006. 'From Long-term Patterns of Seasonal Hunger to Changing Experiences of Everyday Poverty: Northeastern Ghana c. 1930–2000', *Journal of African History* 47(2): 181–205.

Dixon, J. 2009. 'From the Imperial to the Empty Calorie: how Nutrition Relations Underpin Food Regime Transitions', *Agriculture and Human Values* 26(4): 321–333.

Ebbs, J.C. 1991. *The Hidden War*. Edinburgh: The Pentland Press.

Economic Advisory Council. 1939. *Colonial Empire: Part I. Nutrition in the Colonial Empire*. London: His Majesty's Stationery Office.

Eijkman, C. 1896. 'Über den Gaswechsel der Tropenbewohner, speziell mit Bezug auf die Frage von der chemischen Wärmeregulirung', *Archiv für die gesamte Physiologie des Menschen und der Tiere* 64(1–2): 57–78.

Escobar, A. 1988. 'Power and Visibility: Development and the Invention and Management of the Third World', *Cultural Anthropology* 3(4): 428–443.

FAO. 1947. *Rice and Rice Diets: a Nutritional Survey*. Washington: FAO.

FAO. 1950. *Calorie Requirements: Report of the Committee on Calorie Requirements. FAO Nutritional Studies No. 5*. Washington: FAO.

FAO Standing Advisory Committee on Nutrition. 1946. *First Report to the Director-General. Copenhagen, 31 August 1946*. Washington: FAO. FAOA.

Floud, R., A. Gregory and K. Wachter. 1990. *Height, Health and History: Nutritional Status in the United Kingdom, 1750-1980*. Cambridge: Cambridge University Press.

Friedmann, H. and P. McMichael. 1989. 'Agriculture and the State System: The Rise and Decline of National Agricultures, 1870 to the Present', *Sociologia Ruralis* 29(2): 93–117.

Gautier, R. 1941. Letter to Wallace Aykroyd. Geneva, 24 October 1941. LNA.

Hale, S. 1841. *Early American Cookery: The Good Housekeeper*. Boston: Otis.

Haraway, D.J. 1989. *Primate Visions: Gender, Race, and Nature in the World of Modern Science*. New York and London: Routledge.

Hardy, A. 1995. 'Beriberi, Vitamin B1 and World Food Policy, 1925–1970', *Medical History* 39(1): 61–77.

Hargrove, J. 2006. 'History of the Calorie in Nutrition', *Journal of Nutrition* 136(12): 2957–2961.

Harris, J.A. and F.G. Benedict. 1919. *A Biometric Study of Basal Metabolism in Man*. Washington: Carnegie Institution.

Health Organisation of the League of Nations. 1937. *Report of the Intergovernmental Conference of Far-Eastern Countries on Rural Hygiene, Held at Bandoeng, Java, August 3rd to 13th, 1937*. Geneva: League of Nations. LNA.

Health Section of the League of Nations. 1938. *Physiological Needs and Methods of Assessing Health Fitness and Well-Being. Geneva, November 24th, 1938*. Geneva: League of Nations.

Jansen, B.C.P. and W.F. Donath. 1926. 'On the Isolation of the Antiberiberi Vitamin', *Proceedings of the Koninklijke Nederlandse Akademie van Wetenschappen* 29(1): 1390–1400.

Joint FAO/WHO Expert Committee on Nutrition. 1949. *Report on the First Session, Geneva, 24-28 October 1949*. Washington: FAO.

——— 1954. *Requirements for Calories and Nutrients, Provisional Agenda Item 4, Geneva, October 26-November 2*. Rome: FAO. FAOA.

Keys, A. 1930. 'Recent Advances in the Study of Basal Metabolism: Part I', Journal of Nutrition 3(2): 217–228.

——— 1945. 'The Refinement of Metabolic Calculations for Nutritional Purposes and the Problem of Availability', *Journal of Nutrition* 29(1): 81–84.

League of Nations. 1936. *The Problem of Nutrition: Interim Report of the Mixed Committee on the Problem of Nutrition*. Geneva: League of Nations.

Leitch, I. 1942. 'The Evolution of Dietary Standards', *Nutrition Abstracts and Reviews* 11(4): 509–521.

Lewis, J. 2000. *Empire State-Building: War and Welfare in Kenya, 1925-52*. Oxford: James Currey.

Lipphardt, V. 2012. 'Isolates and Crosses in Human Population Genetics; or, a Contextualization of German Race Science', *Current Anthropology* 53(5): 69–82.

Lipton, M. and M. Ravallion. 1995. 'Poverty and Policy', in J. Behrman and T. Srinivasan (eds), *Handbook of Development Economics, Vol. IIIB*. Amsterdam: Elsevier, p. 64.

Mason, E. and F.G. Benedict. 1931. 'Basal Metabolism of South Indian Women', *Indian Journal of Medical Research* 19(1): 75–98.

McCarrison, R. 1924. 'Rice in Relation to Beriberi in India', *British Medical Journal* 1(3297): 414–420.

McNeely, C. 1995. *Constructing the Nation-State: International Organizations and Prescriptive Action*. Westport: Greenwood.

Mellanby, E. 1919. 'An Experimental Investigation on Rickets', *The Lancet* 193(4985): 407–412.

Moleschott, J. 1852. *Der Kreislauf des Lebens: Physiologie Antworten auf Leibigs Chemische Briefe*. Mainz: Victor von Zabern.

Nichols, B. 1994. 'Atwater and USDA Nutrition Research and Service: a Prologue of the Past Century', *Journal of Nutrition* 124(9): 1718–1727.

Orr, J.B. and J. Gilks. 1931. *The Physique and Health of Two African Tribes*. Medical Research Council Special Report Series No. 155. London: His Majesty's Stationery Office.

Packard, R. 2009. 'The History of the Social Determinants of Health in Africa', in H. Cook, S. Bhattacharya and A. Hardy (eds), *History of the Social Determinants of Health: Global Histories, Contemporary Debates*. Andhra Pradesh: Orient Black Swan, pp. 42–77.

Parran, T. and C.K. Lakshmanan. 1948. Report of the Survey Mission to the Far East. Washington, 1 July 1948. FAOA.

Passmore, R. 1950. Letter to Aykroyd. Edinburgh, 23 October 1950. FAOA.

Patil, V. 2008. *Negotiating Decolonization in the United Nations: Politics of Space, Identity, and International Community*. New York: Routledge.

Richet, C. 1885. *Recherches de Calorimétrie: I*. Paris: G. Masson.

Rivers, J. 1996. 'Twentieth Century Growth Theories', PhD Dissertation. Albany: Massey University.

Rubner, M. 1883. 'Über den Einfluss der Körpergrösse auf Stoff- und Kraftwechsel', *Zeitschrift für Biologie* 19(1): 535–562.

——— 1902. *Gesetze des Energieverbrauchs bei der Ernährung*. Leipzig: F. Deuticke.

Ruxin, J. 1996. 'Hunger, Science, and Politics: FAO, WHO, and UNICEF Nutrition Policies, 1945 -1978', PhD Dissertation. London: University College of London.

Said, E.W. 1978. *Orientalism*. New York: Pantheon Books.

Schuknecht, R. 2010. *British Colonial Development Policy After the Second World War: The Case of Sukumaland, Tanganyika*. Münster: LIT Verlag.

Selcer, P. 2012. 'Beyond the Cephalic Index: Negotiating Politics to Produce UNESCO's Scientific Statements on Race', *Current Anthropology* 55(S5): 173–184.

Soekirman, T.I., I. Tarwotjo, I. Jus'at, G. Sumodiningrat and F. Jalal. 1992. *Economic Growth, Equity and Nutritional Improvement in Indonesia*. Geneva: United Nations Administrative Committee on Coordination/Sub-Committee on Nutrition.

Staples, A. 2006. *The Birth of Development: how the World Bank, Food and Agriculture Organization, and World Health Organization Changed the World, 1945-1965*. Kent: The Kent State University Press.

Steckel, R. 2008. 'Heights and Human Welfare: Recent Developments and New Directions', *Explorations in Economic History* 46(1): 1–23.

Steggerda, M. and F.G. Benedict. 1928. 'The Basal Metabolism of some Browns and Blacks in Jamaica', *American Journal of Physiology* 85(1): 621–633.

Tanner, J.M. 1981. *A History of the Study of Human Growth*. Cambridge: Cambridge University Press.

Technical Commission of the Health Committee of the League of Nations. 1936. 'Report on the Physiological Bases of Nutrition', *Quarterly Bulletin of the Health Organization of the League of Nations* 5(3): 391–415.

Technical Commission on Nutrition. 1937a. 'Report by the Technical Commission of Nutrition on the Work of its Third Session', *Bulletin of the Health Organisation of the League of Nations* 7(1): 461–502.

——— 1937b. *Third Meeting of the Commission: London, November 15th, 1937. Point VI of the Provisional Agenda of the Session: Consideration of the Observations Received Regarding the Report on the Physiological Bases of Nutrition. Notes by the Department of Public Health of the Government of British India - A. W. Russell.* Geneva: League of Nations. LNA.

——— 1939a. *Meeting of Members of the Commission and other Nutritional Experts. Note by Dr. W. Aykroyd on The Poor Rice Eater's Diet, 31 July 1939.* Geneva: League of Nations. LNA.

——— 1939b. *Meeting of Members of the Commission and other Nutritional Experts. Note by Dr. A. G. van Veen on Fat Requirements, 31 July 1939.* Geneva: League of Nations. LNA.

——— 1939c. *Nutrition Problems of the Far East, by Dr. A. G. van Veen. 13 March 1939.* Geneva: League of Nations. LNA.

Technical Commission on Nutrition, Sub-Committee on Nutrition in Asia and Tropical Countries. 1938a. *Meeting Held at Geneva, August 22nd-24th, 1938: Summary of Proceedings. Geneva, August, 1938.* Geneva: League of Nations. LNA.

——— 1938b. *Meeting Held at Geneva, September 8th 1938.* Geneva: League of Nations. LNA.

Tilley, H. 2011. *Africa as a Living Laboratory: Empire, Development, and the Problem of Scientific Knowledge, 1870-1950.* Chicago: University of Chicago Press.

Treichler, R. and H. Mitchell. 1941. 'The Influence of Plane of Nutrition and of Environmental Temperature on the Relationship between Basal Metabolism and Endogenous Nitrogen Metabolism Subsequently Determined', *Journal of Nutrition* 1(22): 333–343.

United Nations. 1948. *Universal Declaration of Human Rights, Adopted and Proclaimed by United Nations General Assembly on 10 December 1948.* A/RES/217(III)A.

Vernon, J. 2007. *Hunger: A Modern History.* Cambridge: Belknap Press.

Waterlow, J.C. (ed.). 1955. *Protein Malnutrition: Proceedings of a Conference in Jamaica (1953) Sponsored jointly by the Food and Agriculture Organization of the United Nations (FAO), World Health Organization (WHO), and Josiah Macy Jr. Foundation.* New York: Columbia University Press. FAOA.

Whitmarsh, I. 2008. *Biomedical Ambiguity: Race, Asthma, and the Contested Meaning of Genetic Research in the Caribbean.* Ithaca: Cornell University Press.

Widmer, A. 2012. 'Of Field Encounters and Metropolitan Debates: Research and the Making and Meaning of the Melanesian "Race" During Demographic Decline', *Paideuma* 58(1): 69–93.

Wittgenstein, L. 1953. *Philosophical Investigations.* Oxford: Blackwell.

Worboys, M. 1988. 'The Discovery of Colonial Malnutrition Between the Wars', in D. Arnold (ed.), *Imperial Medicine and Indigenous Societies.* Manchester: Manchester University Press, pp. 208–225.

Wylie, D. 1999. 'Disease, Diet, and Gender: Late Twentieth-Century – Perspectives on Empire', in R. Winks and A. Low (eds), *The Oxford History of the British Empire, Volume V.* Oxford: Oxford University Press, pp. 277–289.

7
The Gender of Nutrition in French West Africa
Military Medicine, Intra-Colonial Marginality and Ethnos Theory in the Making of Malnutrition in Niger

Barbara M. Cooper

In July 1951 the 'Mission Anthropologique' exhibited the fruits of its ongoing research into food and nutrition in French West Africa (Afrique Occidentale Française, hereafter AOF) at the Hall de l'Information in Dakar, Senegal. The exhibition, 'AOF Problème Alimentaire et Nutritionnel' [AOF: The Issue of Food and Nutrition], included a lengthy brochure authored by the local director of the Mission, Médecin-Commandant J.L. Bergouniou; the pamphlet set out general observations about human nutrition as understood in the early 1950s, the nature of foods within the AOF, and a variety of recommended diets depending upon the main food staples (sorghum, rice, millet, root crops and bananas) (Bergouniou 1951).[1] The visual display included maps showing the incidence of goitre across the territory, panels showing different kinds of staple foods and their various strengths and weaknesses, drawings representing ethnic types, and images depicting illnesses associated with nutritional deficiencies. Médecin-Commandant Bergouniou had himself produced most of the watercolours, drawings and paintings that celebrated ethnic types and nutritious fruits, or represented alarming illnesses resulting from poor nutrition. The exhibition was encyclopaedic and visually engaging, and yet as the culmination of six years of collective research it conveyed surprisingly little useful information about how one would actually go about systematically improving the context in which people come to eat what they eat. Given the urgency with which malnutrition in countries such as Niger is debated today, this early effort by the Mission Anthropologique to study food and nutrition in the AOF offers an important window onto how nutrition was approached in the post-war period, and sheds light on why, despite decades of research, the nutritional condition of many in west Africa today remains precarious.

This chapter will consider the unfolding of nutritional science in French West Africa, closing with observations on the reach of such work in the marginal colony of Niger. I will take up this study in three parts. In the first I set out briefly the general context of 'imperial science' in the French milieu, which differed in significant ways from the more familiar case of the British Empire in Africa. In the second I reflect upon the work of the Mission Anthropologique as a whole and how it was conducted. And finally I turn to the specific setting of Niger in the late colonial period, where the problem of malnutrition has historically been particularly acute. Several key themes will be emphasized: the particularity of French colonial science vis-à-vis nutrition due to the imprint of Lamarck and to longstanding internecine struggles within French anthropology; the enduring significance of the highly masculine and military character of French colonial health services; the stark chasm between what appears in reports cast at the scale of French West Africa, and the realities on the ground in territories such as Niger distant from Dakar; the reduction of female significance to the domain of childbirth proper; and the predictability of recourse to explanations based upon 'race' and later 'culture' to account for debilitating malnutrition in the absence of resources sufficient to adequately address the problem.

French 'Imperial Science': Conflicts and Inconsistencies

From the 1920s the French espoused a frank policy of 'faire du noir' [produce blacks] in its West African colonies. The colonial enterprise, it was believed, would succeed only through a dual effort to improve the African population in both 'quantity and quality' while countering population decline in the metropole (Dozon 2008: 45). By the end of the Second World War, overtaxed, understaffed and untrained local administrators had been producing demographic studies of dubious utility for decades in an effort to determine which populations were growing, which were stagnant, and which were in decline (Gervais and Mande 2010). The purpose in principal was to attempt to grasp the cultural and environmental issues affecting population size and quality, and of course to propose concrete remedies. However, as Vincent Bonnecase notes, actual research on food and nutrition in the French territories of West Africa prior to the late 1940s constituted 'a slender corpus' (2009: 152) based upon meagre empirical evidence.

French 'imperial science' from about 1880 through to the 1960s evinced struggles over who should define specialized knowledge about Africa (Sibeud 2002). These debates were bound up in larger struggles over the professionalization of expertise, but also in debates about the nature of civilizational difference and the possibility of fundamental change. On the one hand specialist knowledge consisted of what we would call today physical anthropology – the measuring of bones and bodies both ancient and contemporary, entailing an exhaustive cataloguing of plant and animal life to establish the diacritics of difference. This search for the essential differences between human 'races' had informed the work of Georges Cuvier in the first half of the nineteenth century, work that contributed infamously to the transformation of Sara Baartman from living woman to museum artefact (Crais and Scully 2009). Cuvier's approach to human difference – pitched not at the level of the individual but rather at the level of the 'race' – rejected the possibility of change over time. Cuvier debated evolution vigorously with Jean-Baptiste Lamarck and others, most memorably in his *Eloge* to Lamarck (Appel 1987: 8, 169). Later Paul Broca sought to situate the study of human difference firmly within the natural sciences and evolutionary theory by seeking measurements of human difference in living populations and the fossil record, focusing in particular upon the cranial cavity, which he believed provided an index of intellectual and moral capacity.

As debates over evolutionary theory progressed, unlike scientists in Germany and Great Britain, French scientists tended towards a conception of evolution focused upon the individual rather than the larger group. In this conception of human evolution the key question was what, within the individual organism, could give rise to intra-species variation. Lamarck had (among other things) argued that the individual adapted to a material milieu and proposed that it was possible to pass along those qualities to offspring. The French predilection for what was to become known as neo-Lamarckianism had less to do with Lamarck than to duelling metaphysical commitments on the one hand to a material understanding of causation and on the other to a 'vitalism' insisting upon the autonomy of life forms (Loison 2011). In many ways French anthropology has carried forward literary and philosophical traditions inadequately encompassed by models from the natural sciences (Debaene 2014).

This conception of adaptive evolution was consonant with the notion of the 'rights of man' – French optimism about the capacity of autonomous individuals to be full citizens rejected rigid social classification. This fluid neo-Lamarckian (rather than exclusively Mendelian) understanding of evolution also contributed to France's initial imperial philosophy in Africa, promoting the assimilation of individual Africans to French culture. Assimilationist policy was structured on the assumption that the colonial subject could 'evolve' into a civilized Frenchman (an *évolué*) through the acquisition of French habits of schooling, eating, dressing and speaking.[2] Hygiene, broadly understood, would also play an important part in this civilizing process. As administrators gained experience on the ground in the AOF the notion that passive Africans would be moulded into Frenchmen gave way to a more ambiguous and pragmatic approach in the face of the uncomfortably independent thinking of the *evolués* (Conklin 1997). In the emerging policy of 'association', Africans would develop at their own pace under the tutelage of their own leadership. These leaders would, through association with France, bring the benefits of civilization to their people. Association and its corollary discipline, ethnology, implicitly found value in local cultures and placed less emphasis on the evolution of the individual. This approach had the crucial advantage of facilitating colonial rule by producing indigenous intermediaries (the 'customary chiefs' of each 'tribe' or 'race') through whom the tasks of tax collection and labour recruitment could be executed.

If colonial administrators relied upon the notion of fixed descent-based tribes led by indigenous authorities in order to govern, such fixity was in tension with a more flexible understanding of ethnic difference held by the medical staff overseeing the health of soldiers. French colonial medicine was military medicine, developed to support the needs of the troops known as *Tirailleurs sénégalais*. Practically speaking, as Richard Fogarty and Michael Osborne (2003) argue, French military medicine tended to see racial difference as plastic, influenced by both culture and environment. Military doctors were less interested in fixed difference or capacities than in the comparative pathology of races acclimatized to diverse settings. The key question for them initially had been whether it would be possible for whites to adapt to colonial environments. Gradually as African soldiers recruited in a host of different environments were called upon to enforce French rule in unfamiliar settings, the question of acclimatization became relevant for soldiers of all 'races'. Emphases on fixed biological difference on the one hand and the adaptations of that body through culture on the other were always present to varying degrees within French medical and scientific thinking. Effectively 'race' was slippery and contested; it was also profoundly interwoven with the complexities of imperial engagement.

Thus physical anthropological approaches and cultural ethnological approaches co-existed, resulting in an ongoing struggle to define the nature of expertise on Africa. Scholar-administrators were often less interested in the debates within physical anthropology than in the practical knowledge to be gained through ethnography (the study of language, history, religion and social life of localized cultures) (Sibeud 2002, 2012). Yet the actual health of Africans as opposed to Europeans appears to have garnered little scholarly interest before the late 1920s. A perusal of works on the history of nutritional science in Europe in the early decades of the twentieth century reveals a surprising absence of French scholars (Barona 2010, Kamminga and Cunningham 1995). Prior to the First World War French, British, German and Spanish researchers in tropical medicine had worked closely, in part because of a shared concern for the effects of climate on plants, animals and humans (Osborne 2000). But the rancour of the war estranged French scientists from German, generating increasingly divergent scientific cultures across Europe in the inter-war period (Neill 2012).

In British colonial circles malnutrition was 'discovered' in the inter-war years; in much of Europe, as a result of the stresses of the 1920s, nutrition emerged as a matter of public health (Worboys 1987). By the 1930s a rich literature by anthropologists on nutrition had developed in British territories, in part motivated by anxieties about the capacity of African labour and the possible negative impact of Western influences (Moore and Vaughan 1994; Wiley 2001; Brantley 2002). Meanwhile in the United States a variety of forces conjoined statistics, demography and nutrition into what Nick Cullather memorably refers to as the 'foreign policy of the calorie' (2007). By contrast, Deborah Neill argues that up to the 1920s in French Equatorial Africa, cuisine (a dimension of culture) as opposed to nutrition (the concern of science) marked the French approach to the diet of colonial subjects (2009: 1-28).

Severe famine in France's African territories in 1913–1914 prompted the governor general of the AOF to institute the obligatory construction of 'reserve granaries' to remedy the imagined African characteristics of laziness and improvidence (Bonnecase 2009). This crisis prompted the French Academy of Colonial Sciences to instigate the creation of a technical service on food for the 'natives', but France's budgeting structures required that each colony within the AOF fund its health services through its internally generated revenues, guaranteeing that the necessary resources for determining a proper diet for the region would never materialize, despite periodic prompting from the Ministre des Colonies (Autret 1992: 274). The major developments in colonial nutritional science appear to have largely bypassed French circles – in 1938 one early contributor to the French literature, Henri Labouret, was to express surprise that so little work had yet been accomplished (Conklin 2013: 226).

A comparison of two studies conducted in the 1930s by British and French colonial administrators is instructive in this regard. Both studies offer the results of research into nutrition in the central Sudanic belt. In 1930, W.E. McCulloch, a doctor in the British colonial service in what was then Northern Nigeria, published an 'inquiry in the dietaries' of Hausa and Fulani in urban areas of the territory. The study began with an exhaustive summary of the state of knowledge about nutrition up to that point with a view to providing a convenient compendium for the administrator far from a library. McCulloch then pointed out the importance of highly detailed information about a population's physical condition, its characteristic blood composition, the foods eaten, their chemical composition when cooked, and an assessment of the nutritional components that are deficient or excessive. From all this the careful administrator could, with judicious adjustment, significantly improve the diet necessary for a given population to thrive (1930: 12–13, 69–72). The volume provided a 76-page survey of each of these dimensions for the Hausa-Fulani region, down to a careful assessment of the chemical content of the various kinds of salt favoured in the central Sudan, linking this diet to specific diseases and health problems in the region.

By contrast the 1933 French compendium collected under the direction of Georges Hardy and Charles Richet covered the entire French West African territory as a single unruly and diverse space. A central claim of the volume was that prior to French conquest, the colonial territories suffered from insufficient food, and that it was the conquest itself that had begun to rectify the reckless negligence of the native (1933: 27). François Sorel, who at the time was a doctor for the French colonial troops posted to Indochine, refined this claim: the problem in the AOF was that black eating habits needed to be adjusted to encourage them to continue to consume the very foods they had always eaten, which were well adapted to their constitutions, but to eat them more regularly and in more regular amounts. All of the research upon which Sorel drew had been conducted with a view to determining appropriate rations for soldiers and conscripted labour. Dr Sorel also advocated gradually adapting troops and labourers to the kinds of foods they would encounter in unfamiliar ecological zones, signalling the centrality of mobile troops and obligatory labour across multiple ecological zones to the logic of the medical services of the AOF (Sorel 1933: 169). Henri Labouret (who had been successively a soldier, an administrator and a professor at the Ecole Colonial) offered an overview of the kinds of food produced across the vast territory and expressed confidence that with Western education, the discipline of regular wage labour, and agricultural training a 'real revolution in the economy of black peoples' (1933: 154) that would transform their food consumption was already underway. The volume appeared in the wake of one of the most severe famines in the history of the region. It is hard not to read it as an effort to deflect attention from the broader crisis rather than as a serious contribution to understanding the nutritional needs of the civilian population. What little significant work there was on nutrition in French overseas territories was sponsored not by France, but by the League of Nations (Autret 1992: 275). Such work focused not on the AOF but upon the mandate territories and French Equatorial Africa.[3]

Why, then, the post-war focus among French colonial scientists upon nutrition after decades of neglect? The war in Europe had generated a great deal of unwelcome data on the effects of malnutrition. Much of Europe had been subject to food rationing, and some cities had experienced lengthy blockades creating severe food shortages. Post-war circumstances made it possible to track the health of sub-populations receiving different kinds of rations and supplements (Carpenter 2003d: 3335). Historian Dana Simmons has recently argued that as a result of the war, 'pathological malnutrition entered mainstream medicine' (2008: 181).

With the implementation of the Marshall plan in Europe, the United States and eventually the United Nations Food and Agriculture Organization became increasingly interested in monitoring nutritional health globally, with a particular concern (in an era of growing insistence upon self-government and human rights) that colonial territories not be neglected. UK and US surplus grain and milk also altered the kinds of options imagined to be possible to remedy malnutrition. Rather than meddle with highly complex political economies, it might be possible to simply 'supplement' inadequate diets, while at the same time serving the interests of US and European agriculture (Wylie 2001: 206).

Post-War Nutritional Science: the 'Mission Palès'

This interest in nutrition, led by the United States and the United Nations, put pressure upon France to attend to the nutritional status of its colonized peoples. The first organizational meeting for the Food and Agriculture Organization (FAO) was convened by Franklin D. Roosevelt at Hot Springs, Virginia in 1943, before the liberation of France. André Mayer, a prominent French biologist who had fled France for the United States in 1940, attended the meeting, although in what precise capacity is not clear (Adrian 1999: 19). Mayer had been active in the League of Nations before the war, authoring an important study on the requirements of an adequate daily ration in 1935 for the Bureau of Hygiene (Autret 1992: 276). After the war Mayer continued to be an important figure in the UN and in particular in the FAO; it was his influence that led to the creation of an official French research unit to study nutrition in the AOF. For the first time the French government provided funding for federation-wide research on food and nutrition.

The official French response to the growing international pressure to focus upon food and nutrition was, it seems, ambivalent, in part because an international agenda to guarantee the food security of peoples across the globe would, at the very least, subject French colonial practices to outside scrutiny (Bonnecase 2008: 18, 2011: 111). Furthermore, it might detract from other priorities. As Vincent Bonnecase notes, prior to the Second World War the French took little interest in food except in the context of periodic famine (2011: 30). It would take time for military administrators and medical personnel to adjust to a context in which everyday food consumption conditions would come under scrutiny. The title of the new research effort when it was first formed in 1945 reveals this mix of post-war optimism and French colonial ambivalence: 'L'Organisme d'enquête pour l'étude anthropologique des populations indigènes de l'A.O.F. Alimentation et Nutrition' [roughly, 'The investigative body for the physical anthropological study of indigenous populations of the AOF: Food and Nutrition'].

In the lengthy title 'Food and Nutrition' appeared at the end seemingly as an afterthought, as if merely a subset of a larger, and perhaps more compelling, project. And indeed there was a pre-existing project, for the researcher chosen to head the mission, Dr Leon Palès, had written in 1929 a celebrated thesis on paleopathology using the study of ancient bones to study disease in prehistoric times. In his heart Palès was a paleo-archaeologist, and he eventually went on after the war to do celebrated work on the rock art of Vienne in France, known as the Grotte de la Marche, together with his long-time partner Marie Tassin de Saint-Péreuse. His interest in prehistoric disease meant that he could bridge the not altogether harmonious concerns of physical anthropology and medicine. As the deputy director of the natural history museum in Paris, the famous *Musée de l'homme*, he could oversee the collection of artefacts while conducting the nutritional research called for by the UN.

Palès had been trained at the naval medical school in Bordeaux and at the Ecole d'application du service de santé des troupes coloniales (known as the Pharo) in Marseille, and therefore was firmly located within the tradition of military medicine. Palès had conducted previous anthropometric and epidemiological research (anatomy, physiology and comparative pathology) in French Equatorial Africa among chronically ill railway workers in 1931 (Pépin 2011). Jacques Pépin notes that the fifty careful autopsies Palès conducted in Brazzaville to understand the origins of the wasting condition he referred to as Cachexie du Mayombe are suggestive of some of the earliest cases of AIDS (ibid.: 18–31). Palès later conducted a study among African soldiers stationed in Marseille in 1938 (Mayer 1954: 2).

The choice of Leon Palès to head the mission reflects the profound ways in which colonial medicine in the AOF remained military medicine. Unlike Algeria or to a certain degree Cameroon, French West Africa had relatively few settlers; the aim of the administration was not to cater to a European settler population, nor was it to facilitate the needs of an extractive mining industry. France's relationship towards local societies had been profoundly shaped by the need to recruit soldiers and the concern to characterize the various African 'races' in order to better control and deploy them militarily; France relied upon African soldiers to protect the homeland during the First World War (Lunn 1999) and of course later in the Second World War (Echenberg 1991; Lunn 1999). As a result, a concern to characterize different African 'types' or 'races' continued well after the war and can be seen in the ways in which the 'Mission Palès' made use of convenience samples of *Tirailleurs* and their families.

For Palès' mentor, André Mayer, the challenge facing France in addressing the nutritional needs of the AOF was dual: the complex and varied climate of the West African territories, and the 'obstacle' presented by inscrutable African societies (Mayer 1954: 5). Palès' anthropometric studies held the promise of establishing a knowable fixity in a landscape of moving populations and substantial intermarriage (cf. Scott 1999). As Appadurai has argued, the collection of numerical data according to categories of difference offered colonial powers 'the illusion of bureaucratic control and [became] a key to a colonial imaginary in which countable abstractions [...] created the sense of a controllable indigenous reality' (1996: 117). Since the legibility of the African population was to be effected through the careful tracing of difference, the best way to approach the question of nutrition seemed to be to study the region, in the language of the day, 'race by race'. Thus race would be central to Palès' collection of data, although in keeping with the conceptual fluidity of 'race' within French scientific thinking, the term could mean anything from 'sub-species' (at the physical anthropological end of the spectrum) to ethnic group (for the more ethnologically inclined).

The extraordinary variety of reports and projects eventually produced by the Mission Anthropologique from 1945 to 1955 is intriguing; the research group included public health doctors, haematologists, biochemists, specialists in infectious disease linked to a variety of specialized research centres, epidemiologists, demographers, nutritionists, psychologists and ethnographers (Collignon 2002, 2006). As a result there was often little communication or coherence between the different research elements. Different specialists headed up the anthropological efforts, psychological studies, medical studies of the nutritional status of various populations (soldiers, nursing women, students), chemical analyses of the nutritional content of various foods, the agricultural research, and so on. It was only in 1952 that the mission's name was changed to 'L'Organisme de Recherches sur l'Alimentation et la Nutrition Africaines' (ORANA) [roughly, The Research Organization on African Food and Nutrition], more in keeping with the oft-referred to mandate to respond to the urgings of the FAO. Not long after the publication of its major findings in 1954 (replete with diagrams, charts, assessments of the nutritional content of various foods, and tables indicating the overall nutritional intake of assorted 'races' or ethnic groups) the research agenda of the mission was to all intents and purposes disbanded. A new phase of implementation was, in principle, to succeed it.[4] In practice, with the advent of independence, ORANA focused more narrowly on nutrition in Senegal and lost its broader regional character.

Much of the initial work of the Mission Anthropologique continued a tradition within physical anthropology known as 'raciology' (*raciologie*) in which body measurements and the administration of intelligence tests were taken with a view to setting out the population level differences between different physical types or 'races'. The project therefore assumed in advance that there were racial differences, and then went about characterizing and literally mapping them. Bonnecase argues that for Palès these measurements served as a means of gaining a more rigorous taxonomy, a seemingly objective measure, of the degree of racial intermixing of 'Negros', 'Ethiopians' and 'Arabs', in order to skirt the 'false' self-designations of populations in this extremely hybrid region (2011: 128).

However, race was not the only frame through which the work of the mission progressed. The other defining frame was the assessment of male labour capacity. Palès' previous work had examined the nutritional status of 400 African soldiers in Marseille (Palès and de Saint-Péreuse 1949: 5). Soldiers have often presented social science and medical researchers with a useful 'convenience sample'.[5] The medical interest in soldiers in the AOF, however, as in other parts of the colonial world, was also concerned with their capacity to perform a particular kind of labour: the goal was to have effective soldiers, not simply healthy subjects. By extension the aim was to find ways to improve the productive capacity of African male labour more broadly. In the words of André Mayer, 'One can't bring out the potential of a corner of the world without bringing out the potential in the men who occupy it, without encouraging them to develop to the full their capacity for thought and action' (1954: 8). The Mission Anthropologique bore the strong imprint of France's military presence in Africa, as well as the historical legacy of a medical system designed to preserve the health of the troops and only secondarily to address the African civilian population.

Palès knew, of course, that the sample he had from among soldiers would not be representative of the population as a whole. In effect he used the relatively robust and well-fed male soldiers as a kind of ideal 'norm' against which to measure the actual physical and nutritional shortfalls among civilian members of the soldiers' respective ethnic groups. The temptation previously had been to take soldiers as convenient representatives of their ethnic 'types'. Instead Palès inventively used the military population to try to gain insight into the relatively poor condition of civilians of the same ethnicity who had not benefited from military rations: 'the soldier becomes thereby physically an index of what could be but is not […] it is he who gives us the measure of how much change perceptible improvements in food and essential nutrition can bring about in African populations' (Palès and de Saint-Péreuse 1953: 20). The question of how to determine appropriate caloric and nutrient intake was a pressing concern for all the colonial powers but it posed the problem of whether a European standard of consumption should be the ideal;[6] Palès' approach had the virtue of establishing an 'attainable' baseline (the rations of soldiers).

To do so he and his partner Marie Tassin de Saint Péreuse took 11,344 measurements of adult male military and civilian subjects to establish how robust they were (Palès and de Saint-Péreuse 1953: 20). He also took a host of measurements whose relevance to nutritional status is less obvious (Palès 1948: 8–9).[7] The mission took, for example, 2,000 photographs and 1,000 x-rays of African heads in profile to measure prognathism – commonly used in ranking races relative to one another along a civilizational spectrum. The mission also undertook dozens of autopsies and dissections to study the soft tissue of black Africans, work that would be as useful for studying infectious disease and cancer as for studying nutrition.

Despite the credible and in some ways rather forward looking agenda to generate a local sense of the nutritional 'norm', Palès' research clearly continued to engage with a positive social hygiene tradition aimed at improving the races 'both in quantity and in quality' that was increasingly out of favour with the rise of the National Socialists (Conklin 2002). An emphasis upon material culture and the physical body nevertheless still dominated within the *Musée de l'homme*, where Palès served as assistant director and for which he collected materials while in the AOF. We see the strains and tensions within French anthropology at work. Certainly, in 1952 French cultural anthropologist Claude Lévi-Strauss argued publicly against any teleological notion of progress and emphasized 'the absurdity of claiming that one culture is superior to another' (41) given the importance of cultural borrowing throughout history.[8] The shift in France from physical to cultural anthropology initiated under Marcel Mauss and decisively established after 1950 coincided with UNESCO's post-war efforts to counter global racism (Conklin 2013: 283–338). In some ways the contradictions within the Mission Anthropologique contributed to this shift.

Among the important discoveries of Palès' comparative work was that hypoglycaemia, which he had hypothesized to be the 'normal' condition of the African, did not exist among soldiers on a good diet – as Palès was to note, 'in this respect, human physiology is one' (cited in Bonnecase 2011: 129). The mission also mapped the incidence of various diseases long associated with nutritional deficits such as goitre and night blindness, as well as others less clearly related to nutrition but of interest in the study of race such as sickle-cell and cancer. It is this disease mapping that appears to have most interested Palès personally (Ferembach 1988: 298). Because in the end diseases tended to map out in ways that coincided far more with diet than with phenotype, the work had the unexpected outcome of undercutting the very logic of race, at least insofar as it referred to physiology as opposed to culture. In effect the very failure of Palès' disease mapping to support the notion that race alone could explain differences in health status or susceptibility to disease contributed to the undermining of the notion of race as fixed and of biology as destiny.

Nowhere in the ORANA materials do we see evidence of the circular reasoning Galluzzi Bizzo notes in chapter 6 (small people simply require fewer calories, which is why they eat less) – in keeping with the global post-war trends there was a clear recognition within ORANA that the diet across the AOF was regularly inadequate – for the stark contrast between the well-being of civilian and military 'natives' precluded any such thinking. What the 1954 report revealed most starkly was the very deep need for more sustained attention to the dynamics producing malnutrition in the AOF. The final report found diets deficient in calories and poorly balanced all across the territories. It accounted for this inadequacy in a variety of ways (poor purchasing power, seasonal shortfalls, cultural habits, a mixture of the above), but the overall picture was one of an entire region in which people struggled to meet even the most basic caloric needs, much less a varied and balanced diet.

Having established inadvertently that 'race' had little utility in explaining either disease or malnutrition, two not entirely congruent explanations for malnutrition were forwarded within ORANA in this period, although not necessarily by the same researchers. One account focused on the 'native nutritional errors' or the poor eating practices that purportedly contributed to malnutrition. For example, the 'natives' chose to grow manioc, which is potentially toxic and poor in nitrogen content. Their diet was monotonous, rarely enlivened with beans or roots (although the nutritional richness of condiments was often remarked upon). Native diets were also poor in animal products such as milk and meat, products that are:

> indispensable for the nursing mother and the child at weaning – it is at that moment that ravages of gastro enteritis are felt and that kwashiorkor appears among small children whose livers begin to fail, a problem that can sometimes be mortal, but for those who survive can be the beginning of a cirrhosis that will cause their death well before they reach 40 (Bergouniou 1951: 15).[9]

In an echo of the cultural racism Diana Wylie (2001) describes in her rich study of discourses surrounding malnutrition in South Africa, the impulse among French nutrition specialists was to see the 'failure' of natives to eat more meat as the product of an irrational attachment to cattle as capital rather than as a productive investment to be exploited: the challenge would be to 'convey to people that herds must be exploited and should not remain unproductive capital' (Bergouniou 1951: 15).

The other consistent way of accounting for poor nutrition also attributed it to poor choices on the part of the native, but it nevertheless acknowledged the possibility that Western intrusion might have in various ways contaminated, or rather diverted, local eating habits. In the wake of the war the dangers of Western intrusion were compounded by the seductions of other movements – implicitly pan Africanism, labour movements and communism, generally associated with urban settings (Mayer 1954: 10). It would be the task of the Mission to address the ways in which the lure of urban life, of modernity, and of competing social movements might generate a weakened and unhealthy work force. Once again resonating with earlier work in Anglo territories on the rising fear of 'detribalization' and its purported link with the poor health of natives in southern Africa (Moore and Vaughan 1994), Palès was to suggest that with Westernization and modernization, Africans were losing 'ancestral knowledge' of rurally based and ethnically inflected food practices most visible not in staple foods but in condiments and snacks such as wild fruits (Palès and de Saint-Péreuse 1953: 22).

Implicit in Palès and his team's admiration for highly nutritious indigenous foods was a criticism of Africans' abandonment of this rich diet in the pursuit of the status that comes with a more Westernized lifestyle. The role of the Mission, then, was not to discover something new, but to persuade the native to return to the riches of Africa's natural wealth: 'West Africa, the land of malnourishment, is also the land of the richest sources of vitamins: vitamin A (and D) in shark liver oil, vitamin B 1 from the pomme du Cayor, vitamin C from the Detarium tree and from the Bauhinia' (ibid.: 27). This approach also had the immense advantage of costing very little, placing the burden upon Africans to return to the past. Despite many references to the ways in which the temptations of Western foods, clothing and entertainments detracted from Africans' ability to nourish themselves appropriately, no evidence or examples were offered to substantiate this claim (Palès 1954: 225–226).

A particularly unfortunate effect of the segregated quality of the research components of the mission was that those most deeply engaged in the nutrition and disease research were professionals with little grasp of agricultural economics. It was all very well to hearken to a romanticized past of rural plenty, but such a vision was rather out of keeping with the observations of the agronomists attuned to local farmers preoccupied with problems of sustaining a fragile and increasingly stretched land base. Some of the more trenchant observations in Palès' final report came from his own careful reading of the most recent work on tropical agriculture, which was highly critical of the soil degradation in Africa under colonial rule.[10] Palès had as a result of such studies a deep appreciation of the technical limitations facing African farmers given the poor soils, the relatively small labour force, and the limited availability of land. Very much in contrast to the discourse of the negligent African farmer common in settler colonies, Palès wrote, if not admiringly, at least with some understanding about the extensive farming methods typical in the region, the logic of burning grasses and branches to return nutrients to the soil, shifting cultivation and the extremely lengthy fallows required to sustain these tropical soils. Of crop rotation and mixed cultivation he was to remark, 'we have little to teach them in this area' (1954: 27).

Palès, in the final report, expressed his fear that this shifting agriculture was, in effect, a luxury, sustainable only so long as the population remained sparse and land plentiful. By the early 1950s the earlier colonial concern to increase the size of the African population had given way to unease about the implications of a population that was growing with the regression of major epidemic disease. With the expansion of education farming was becoming less appealing, so that fewer producers were feeding larger numbers of urban consumers. Mechanized agriculture and the use of fertilizer – the intensified approach of European farming – was unknown and in some ways impracticable here. Far more likely, Palès feared, would be an ever more extensive agriculture at the expense of fallow periods, potentially aggravated by the promotion of animal traction (ibid.: 28).

However, this level of nuance and reflection on the dilemmas of tropical agriculture is utterly absent from the shorter, and presumably more widely distributed, brochure produced for the 1951 exhibition 'AOF Problème Alimentaire et Nutritionnel'. The exhibition and brochure were overseen by Médecin Commandant J.L. Bergouniou, who replaced Palès when he returned to France in 1951. Bergouniou was at that time the health officer for the Casamance region of Senegal. In the rare practical recommendations in the brochure, the nutrition specialist seemed to sidestep the realities of political-economy and soil science, insisting that the native 'ought' to grow more grain by using more modern methods: employing DDT to protect legumes, growing soy, and consuming his herd (1951: 15–16). The food problem, in his view, was one of archaic agriculture – a cultural explanation. It is hard to know how many people visited the 1951 exhibition, but it seems probable that Bergouniou's exhibition and handsome 23-page brochure had more popular impact than the more limited administrative circulation of Palès' lengthy final report.

It was only after Palès' departure, under médecin-Lieutenant-colonel André Raoult that ORANA fully cast off its anthropological legacy (Adrian 1999: 20). In another major change, the Loi-cadre of 23 June 1956 granted each of the colonies within the federation of the AOF a semi-autonomous status that ultimately undermined the financing and utility of an institution such as ORANA. Consequently, ORANA effectively shifted from being the seat of research into food and nutrition for all of the AOF, to addressing such concerns in Senegal exclusively. The work of translating the broader findings into practical applications in other colonies, notably Niger, fell by the wayside.

Bergouniou was a central figure in the post-war development of a public health infrastructure in Senegal, and he became known for a number of co-authored studies on the impact of kwashiorkor on the liver that were done under the auspices of ORANA. Interestingly the Mission, to my knowledge, never produced an actual map of kwashiorkor across the AOF. Perhaps childhood malnutrition, understood to be a form of protein-calorie malnutrition due to 'weaning syndrome', was so common that it would have made little sense to try to map it against 'race'. Perhaps Bergouniou was not particularly interested in Palès' mapping in the first place.

As a nosological entity kwashiorkor was only coming into visibility in the early 1950s. The Mission researchers, including Bergouniou, shared their scattered research findings to a larger World Health Organization study, *Kwashiorkor in Africa*, co-authored by J.F. Brock and M. Autret. Against the backdrop of studies done across the continent, the disease appeared to be relatively severe in French West Africa, involving lengthy bouts of diarrhoea. The co-authors suspected that this particularity had to do with a vicious cycle of gastro-intestinal disorders and dietary deficiency. The relative lack of protein in women's diets appeared potentially significant, as did the regional tendency to use enemas to prompt bowel movements in small children, and the spicy diet offered to children during weaning (Brock and Autret 1952: 27, 37, 43). While the work of the Mission came into play in the study, the impression one gets from the volume overall is that the most important work on kwashiorkor was being done in Uganda, Kenya and the Belgian colonies of Congo and Ruanda-Urundi.

Senegalese historian Mor Ndao has characterized French interventions in the domain of nutrition and women's and children's health in colonial Senegal as a 'succession of flimsy short term solutions made on the spot, with neither coherence nor efficacy' (2008: 207). Health interventions to improve maternal and child outcomes ran up against the perennial problems of scarce resources as well as local hesitation in the face of Western medical treatment, particularly by women who refused to be subjected to examination by male doctors (Ndao 2005: 5–6). As Ndao points out, much of the investment in improving childhood nutrition was focused upon the 'useful portion of Sénégal' (le Sénégal utile) – the westernmost portion of that colony, particularly the agriculturally rich region of Casamance, understood to be its economic engine. Much less investment made its way further east in Senegal, where undoubtedly nutritional circumstances were far more challenging.

Gendering Nutritional Science: Female Knowledge and Male Labour

If the impact of the Mission's work in Senegal was limited and ambiguous, Niger was very far to the east of 'useful Senegal'. Indeed, it was so far east that Palès' extraordinarily detailed ethno-epidemiologic maps never covered the whole of the colony, if they covered it at all. Niger was so marginal to the AOF as to barely merit representation. Generally, the studies touched in a sketchy way upon Niamey and environs before heading back to more useful territory. The question then arises: what impact could the Mission Anthropologique in fact have had on nutrition in a colony such as Niger? The problem is really twofold. First, even at its best the approach to nutrition was gendered in ways that meant that, as Ndao points out, it was unlikely to hit its target. Second, the Mission made few inroads into territories to the east of Senegal and, after Palès' departure, none at all.

The mission's emphasis on maximizing potential male labour through interventions tailored to different regions and ethnic groups rendered the needs of women and children less visible. The lack of attention paid by the mission to women and children becomes clearer when one assesses the very modest attempts of the research team to pursue two absolutely critical issues – childhood growth and nutrition in pregnancy. Unfortunately, despite an initial effort to take careful measurements of children's growth – the sort of intervention that could in the long run have had an enormous impact on childhood health work in the region – the mission soon set aside the subject. It appeared too difficult, given the extremely poor development of birth and death registrations across the AOF and the inaccuracy of the ages offered by potential subjects: 'there is therefore a fundamental element of error that prohibits any scientific undertaking' (Palès 1948: 11).[11] Unlike inter-war Angola, where doctors engaged in an imperfect but nevertheless impressive effort in medical demography through oral interviews and on site registrations of births and deaths by medical sector,[12] the personnel of the Mission Anthropologique did not succeed in adopting a similar work-around.

Thwarted in taking a snapshot of many ages at one moment, it doesn't seem to have occurred to the team that a longitudinal approach following children from birth could be more useful in any case. Such a longitudinal approach might have blurred too readily into the kind of medical focus on individual cases from which the anthropological team – determined to trace ethnicities and populations as a whole – hoped to distance itself. The unclear time horizon for the Mission may also have made it difficult to attempt a longitudinal study.

Similarly, after a single early study of blood work among a convenience sample of soldiers, civilian men, and women in Dakar, all attention to women as potentially distinct sub-populations was completely dropped. It was dropped precisely because the women most accessible for scientific study were those attending a post-natal clinic, a relatively convenient sample. However, 90 per cent of the women were found to have a Vitamin A deficiency as compared with only 8.3 per cent of the soldiers and 46 per cent of the civilian men. Women's pregnancy and childbirth rendered their blood samples difficult to interpret: were the women's diets poorer than the men's, or were they simply losing their nutrients temporarily to the foetus? The remedy appeared to be obvious, to have women build up their reserves through 'the natural sources available in Africa' but no one went on to study why parturient women were not already eating these natural sources of iron, or whether they managed to bounce back while still nursing (Palès 1948: 17–19). A study of some of the women's breast milk revealed that it carried sufficient quantities of vitamin A for the nursing baby. Not surprisingly, not all of the women in the study were eager to give up samples of their precious milk and some had a habit of disappearing before the study of whether shark's oil improved the women's milk could be completed: 'it is useful to emphasize that women generally go out only on the eighth day after giving birth, and that because of various superstitions, we ran up against difficulties in the course of taking milk samples. In particular there is a belief that an infant will die if his mother gives his milk to another child' (Auffret and Tanguy 1947: 230). Rather than explore the meanings of milk and why it was so important to women to protect it, the researchers turned to other issues.

Apart from this single, ambiguous and evidently inconclusive study, no other published research by the Mission focused specifically on women. As is so often the case in the history of medicine, women were sidelined because their changeable and complex bodies were inconvenient and did not readily provide a 'norm' against which deviations could be measured. And, equally importantly, women could not be reduced to measurable ethnic traits if there were no female soldiers to serve as index. Furthermore, women's belief systems touching on childbirth, nursing and motherhood could be complex, requiring a degree of cultural agility that the researchers did not have and that the male African assistants could not surmount. In effect, all women's bodies were so exceptional as to be brushed aside as irrelevant to the question of shoring up male labour.[13]

And yet women were and are absolutely central to agricultural production, food transformation, cooking, child-feeding, and so on – indeed it could be argued that any study of food and nutrition ought to have placed women at the centre. A closer perusal of the Mission's work suggests that, in fact, the researchers did rely extremely heavily on the expertise of indigenous women. To assess the nutritional value of the diet in the western Sudan the researchers (or more likely, their African subalterns, the drivers and translators crucial to the execution of the project) consulted the wives of *Tirailleurs* (soldiers) to establish the weight of the components of a typical meal for a couple and a young child; they purchased cooked foods produced by women at weekly markets to find out what the nutritional value of snacks might be; they noted the diets of families of fisher folk along the waterways who seemed in particularly good health (Palès 1954: 160, Planche XXV).

In seeking out useful informants Palès forthrightly confessed that the ideal circumstance would be to find a polygamous chief who spoke French whose wives were open to assisting:

> In Futa Jalon, one day, we were explaining the purpose of our data collection to an important Fulani chief. 'This is women's business,' he replied, and we were taken into the presence of his first wife who was herself overseeing the grain grinding in a large room. Within a few moments, with perfect clarity and precision, we were fully advised on the preparation of rice, fonio, maize, and the various sauces that accompany them. All that is left is to identify on the site the scientific names of the vernacular plants she had just indicated (1954: 106).[14]

However, women were imagined neither to be the major recipients of the benefits of improved nutrition, nor to be a labour force or source of knowledge worth protecting or cultivating.

Most puzzlingly, women's role as the producers of food and as contributors to a complex food production system that went well beyond the staple millet, rice and manioc fields was ignored even when the researchers appear to have been aware of it. Palès noted that one of the problems with local agriculture was the paucity of materials that could be used to fertilize the soils. He commented dismissively that 'as for animal manure and human waste, they serve only to enrich the gardens adjacent to the huts, land left to the care of women' (ibid.: 27). Given the high nutritional value of the condiments and sauces produced from these gardens, this neglect is puzzling.

In effect the spatio-temporal schema generated by the Mission to capture the farming systems across the region focused almost exclusively upon the staple crop fields so often controlled by men regardless of the importance of female labour on the fields. Yet women did farm throughout the region, whether in their own fields or in gardens closer to the house. Because women often used their production to transform into foods, often highly nutritious snacks sold door to door or in the market, a significant dimension of the story appears to have been overlooked. The spatio-temporal schema appeared at first glance to be gender neutral, but in reality reflected the preoccupations of a male household head. It is possible that the rigid division of labour by sex noted by Palès made it relatively difficult for the male researchers to grasp the nature of women's contributions to the farming system – more visible to them were tasks marked as feminine such as collecting water and pounding grain. While they had to have been aware of women's food processing and marketing activities (after all they sampled women's wares in the market), the studies don't attempt to make sense of how women acquired the elements, what they did with the profits, or how those kinds of foods (unlike the exceptional family meals cooked by wives of *Tirailleurs*) circulated. The 'social relations of consumption and exchange' emphasized by Henrietta Moore and Meghan Vaughan in their study of colonial work on nutrition in Zambia fell very much outside the purview of that schema (1994: 47).

Ethnos Theory: Faulty Feeding and the Neglect of Political Economy

As Diana Wylie notes, one tendency within colonial nutritional science was to resort to 'ethnos theory' to account for the poor health of colonial subjects (2001: 218). The theory went that the balanced diet of the untouched native had been disrupted with the advent of European tastes. In the context of South Africa, this way of making sense of poor nutrition required imagining rural and urban spaces in separate and disjointed frames. The problem of the 'urban native' was his detribalized diet. The problem of the rural farmer was his poor technique and irrational investment in cattle. Rather than interpret embarrassing statistics revealing the magnitude of kwashiorkor in light of land laws and labour practices, it was convenient to re-read such data in light of race and culture (ibid.: 222–224).

In a parallel manner French colonial science read the space of the AOF in ways that documented bodies as fundamentally ethnic, with little attention to gender, social status or history. The normal body was a male body. Although preventive rather than purely curative medicine was recognized as important, the focus of most interventions touching on nutrition was only African workers and soldiers (largely male) and their families (Domergue-Cloarec 1997: 1233).[15] Of course, prior to the late 1950s health professionals were not aware of the importance of nutrition-immunity-infection interactions, so the significance of nutrition interventions for addressing infectious diseases was not yet well known (Keusch 2003).

Nevertheless, by the late 1940s scientists were increasingly attentive to the health effects of protein shortfalls, although they did not entirely understand how amino acids actually worked. In part as a result of work in the late 1940s and 1950s collecting data on the health of children, kwashiorkor had been identified as a specific problem thought at that time to be largely the result of inadequate protein consumption. By the 1960s this emphasis on protein had led to the declaration of a 'world protein problem' (Carpenter 2003d: 3337).

Oddly, the colonial discussions one finds on prevention often ignore the implications of the introduction of cash cropping for nutrition; the explosion of peanut production for export, for example, was not explored as one significant dimension of nutritional well-being. In particular, no attention was given to how such shifts might have affected women's nutrition or that of children. Women came into visibility only in the context of medical centres devoted to the health of pregnant women and small children, centres which, while valuable, were few on the ground and of ambiguous impact.[16] While valuable, they could never adequately address the problem of nutrition in the AOF. Furthermore, the attention paid to women exclusively in the context of infant and maternal health tended to reduce the female body to the reproductive body, occluding the significance of women in production and sidestepping questions of wealth distribution.

For the historian the Mission Palès documents are particularly frustrating because by neglecting to see African food production systems (as opposed to consumption) as having undergone significant change, the materials give us no tools to ascertain whether the incidence of malnutrition had gone up, whether occasional food scarcity was relevant, how taxation and cash cropping might have had an impact on nutritional health, and how shifting demands upon male and female labour would have shaped food consumption. Monica van Beusekom's (2002) work on the massive irrigation project in Soudan (contemporary Mali), known as the Office du Niger, has underscored the importance of African food needs and preferences to the well-being of African farmers under the heavy labour and taxation demands of the French empire as cash cropping gained primacy. Women in the vast irrigation scheme were, according to Laura Ann Twagira (2013), fundamental to the transformation of this 'modern' landscape into a foodscape amenable to supporting the nutritional needs of the populations forcibly settled there. The researchers of the Mission Palès made no attempt to place the nutritional deficiencies of the AOF population in the context of this kind of shifting political economy.

Women's Work and the Condition of Children in Niger

Nevertheless, from reports produced by the Service de Santé of the Colonie du Niger it is quite clear that from very early on health officers in that colony were aware of the terrible toll infectious diseases were taking on small children; in fact the earliest of the documents on Niger collected at the Institut de médecine tropicale du Service de santé des armées (IMTSSA) was a study of a devastating meningitis epidemic in 1937–1938 that hit children from two to fifteen years old particularly hard (Pauliac 1938). While that initial report did not focus upon nutrition, it did signal the importance of problems related to children's health.

The collection of a series of demographic studies across the Niger territory suggested that in some regions, despite quite high birth rates, population size remained stagnant (Saliceti 1941: 94–95). A variety of reasons were adduced to account for this pattern, including the purported effects of polygamy (conjoining impotent elderly wealthy men with multiple nubile young women), damage caused by premature commencement of sexual activity for girls married young, high sterility rates in some regions, and miscarriage caused by syphilis (ibid.: 95–96). Another factor proposed was the premature aging of women who were overworked far into pregnancy (idem 1943: 112). Women were also assumed to engage in induced abortion, which then led to sterility (ibid.: 113).

Nevertheless, health officers were also aware that infant mortality was quite high and that infertility and sterility did not fully account for the low population growth in many regions (idem 1941: 97). Le Médecin Lt. Colonel Saliceti had in the early 1940s elaborated a notion of '*débilité congenital*' (congenital infirmity) as one way of accounting for this high rate of infant mortality. The theory was that, as a result of generally poor hygiene, close birth spacing, and the overwork of women, children were born with a kind of physical retardation that contributed to their vulnerability to illness (idem 1943: 114). The notion dovetailed nicely with the neo-Lamarckian tendencies of French social hygiene at the time. 'Reprehensible customs' were, he argued, also important contributing factors, including washing babies too often in the cold, failing to dress them properly, and permitting them to become filthy once they were crawling (idem 1941: 97). Some practices of indigenous birth attendants were seen as particularly dangerous, such as cutting the umbilical cord with dirty instruments, and using pre-masticated juices as astringents, leading to tetanus and septicaemia (idem 1943: 115).

However, infant feeding practices were particularly noted and were critiqued with a degree of detail worth exploring:

> Cercle de Fada-n-Gourma (1940): The gourmantché infant is also nursed at the breast. Additionally twice daily there is a forcible feeding of an infusion of tamarind and 'sesseguéri' to 'strengthen the child'. If the mother should not happen to produce milk, a relative – but only a relative – will place the child at her breast. If this isn't possible then the child is only fed the herbal infusions. The child is nursed for at least one year, often longer. At that time suddenly [*brutalement*] the child is given millet paste with a lightly spiced sauce. This shift, without any transition, from milk to adult food obviously can't occur without occasionally severe gastro intestinal problems (idem 1941: 97).
>
> Cercle de Zinder (1942): The child is exclusively breastfed for a year or until the mother becomes pregnant again. From one year the feeding is accompanied by an uncooked millet paste to which a bit of sour milk is added. The mother continues to offer milk for up to 21 months for boys, 20 for girls. The child is placed at the breast at all hours of the day or night, without any limit. Nevertheless digestive problems are rare, as mother's milk is the best protection against gastro enteritis. The sour milk added to the millet flour produces lactic acid which also serves as an effective intestinal antiseptic (idem 1943: 106).
>
> Cercle de Tanout (1942): Early and brusque weaning or solid foods fed too early and in too great quantity (millet paste from the sixth month) are the cause of many cases of gastro enteritis, which, when aggravated by the cold, becomes a major cause of mortality (ibid.: 115).

For Saliceti, a number of 'causes' accounted for the low population growth rates across the Cercles of Niger, including syphilis, gonorrhoea and malaria. But most important were a variety of customs that produced the 'feeble child' (*enfant débile*), who could not withstand illness and the climate (ibid.: 117). The goal, therefore, was to address these unfortunate customs and in particular those related to nursing, infant feeding and weaning.

A number of features of the attention to infant feeding are striking. Women's 'failure' to schedule infants' nursing perplexed medical officials even though in some instances (as in Zinder) it did not appear to cause problems. Saliceti objected to feeding children adult foods at six months in Tanout, but was equally perturbed by the Gourmantché practice of nursing for a full year. In Zinder the addition of fermented milk to infant foods was noted approvingly, whereas elsewhere it is seen as a pernicious local custom (Morvan 1946: 53). In other words, local practices appear to have been objectionable in regions with perceived low population growth rates and commendable where the scanty available evidence appeared to indicate growth. Altering these seemingly problematic behaviours would require the intervention of a female health professional. In general, it seemed, in order to address the health of women and children it would be first necessary to address the problem of generating a cadre of educated women (Comité de la Journée des mères de familles nombreuses 1920–1923; Brévié 1935).

By 1939 some attempt was being made in Zinder to focus upon women and children's health and hygiene; a trained nurse was hired, presumably recruited from among the small number of African *infirmières visiteuses* (visiting nurses) trained at the Ecole de médicine in Dakar from 1925 to 1939. The school had begun training *sage-femmes* (professional midwives) in 1918, recruiting female students largely from Dahomey, Senegal, Soudan and Côte d'Ivoire. Students who were not successful in passing into the advanced courses for midwives could, from 1928, graduate with a certificate authorizing them to serve as 'missionaries of hygiene'. In practice this outreach work meant tracking women's childcare practices in their homes and metaphorically hounding them (they were to be *rabbateuses*) until they made use of the available colonial health services (Barthélémy 2010: 30). Niger sent very few young women to be trained in the medical school because the educational system in the colony was too rudimentary to yield many educated girls. Niger was regarded by successful midwives as an undesirable posting – the presence of a 'nurse' rather than a 'midwife' probably reflects the difficulty of recruiting the more highly trained *sage-femmes*.

This unnamed nurse appears to have taken a rather rigid and insensitive approach that alienated women, very much as Ndao suggests occurred in Senegal. Le Médecin Lieutenant-colonel Guillaume commented in his annual report in 1940 that although the nurse appeared quite qualified on paper, 'her rather dictatorial character unfortunately prevents her from entirely gaining the confidence of the natives' (Guillaume 1940: 36). This may have been less a character fault on the nurse's part than the outcome of poorly conceived training and paternalistic objectives that were ill suited to the realities on the ground. The position of *infirmière visiteuse* proved less useful than initially hoped for throughout the AOF, and from 1939 the Dakar medical school ceased training for this disappointing category of health worker (Barthélémy 2010: 30).

The concentration of the scarce trained women in their colonies of origin meant that the colonies of the Sahel, including Niger, had as late as 1942 no real medical services to speak of for women and children (ibid.: 178). Interestingly, the only consistently successful effort to reach out directly to women with children in this period appears to have been taken up by the Red Cross and other similar voluntary groups run by colonial women.[17] Across the AOF, as part of the French effort to promote childbirth, such groups sponsored beautiful baby contests and baby weighing events – the emphasis was more upon promoting population growth than upon nutrition (Comité de la Journée des mères de familles nombreuses 1920–1923). Beginning in 1936, French women (largely the spouses of administrators) in the capital (Niamey) had undertaken weekly baby clinics with a view to weighing infants so that if they appeared to be doing poorly their mothers could be encouraged to take them to a dispensary.

Baby weighing was encouraged through the giving of small gifts of clothing, millet, soap and milk to the mothers who attended (Saliceti 1941: 76). The modest and essentially charitable activities of the Red Cross in the capital, for as long as they continued, were featured centrally in annual medical reports, as if to absolve the Service de Santé of any obligation to address proactively maternal and infant health across the vast territory of Niger: 'the activities of this charity meet very nearly all the needs relative to the protection of small children' (Saliceti 1943). The activities in Niger were so paltry that the health reports do not make any effort to report numbers of consultations, mentioning instead the distribution of food and soap to 'needy mothers' (Bonnecase 2011: 50). Unfortunately, by 1950 the Red Cross in this form was no longer functioning in Niger, leaving a yawning gap in the outreach to women and children (Lorre 1951: 21). For whatever reason, French colonial wives in Niger were devoting their energies elsewhere.

Within the medical corps proper, the authoritarian military medical culture prevailed over these modest charitable approaches. Military doctors often turned to culture to explain the failure of African populations to adhere to desired health practices.[18] Medical staff, although too few in number to accomplish much on the ground, were not content to let women make decisions on their own about prenatal care, maternity birth, and postnatal care. 'Traditional' birth attendants were to be enticed to draw parturient women to the medical facilities. Certainly, some positive inducements such as gifts of baby clothing were offered. Intriguingly, women in Maradi were briefly urged to stay at the maternity a full week before they were expected to give birth, with the costs supported in principal by the Société de Prévoyance (Morvan 1946: 55). While this might at first blush appear quite generous, when one reflects upon the psychological effect on a pregnant woman of abandoning her other children for that long it is less clear that such 'generosity' would have been welcome.

That such efforts actually entailed a degree of coercion becomes even clearer in one report on maternity practices in Zinder. Indigenous women's 'natural repugnance' to seeking prenatal medical care was at times overcome through 'more effective arguments' such as the possibility of administrative sanctions. Morvan remarked in 1947:

> When it comes to giving birth women exhibit a passive resistance that is very difficult to overcome. At one point a system of punishments was applied

> here: women who gave birth at home without immediately informing the maternity from the beginning of labour were transported to the maternity in the subsequent days and remained there for some eight days for care. This practice provoked incidents and was eliminated. But mere persuasion and obligating the traditional birth attendants to call the midwife as soon as they are called for a birth are not sufficient to overcome resistance due to custom, religion and atavism (52).

Far from seeking out women's knowledge of childrearing, nutrition and health, colonial medical officers in Niger treated women as criminals. Not surprisingly, in Zinder women tended to evade the control of the health services. Despite efforts to attract women to give birth in clinics, such as raising the bonus offered to local birth attendants for bringing women to the maternity, attendance at maternities in the late 1940s actually went down rather than up (Kervingant 1948: 31). The medical staff lamented the passing of required gynaecological exams for prostitutes once prostitution was declared illegal in 1946 (idem 1949: 41). Such authoritarian approaches were very much part of the medical culture, accounting to a large degree for its inability to engage productively with women and their children.

Under the circumstances the intervention of the anthropological and nutritional research staff of the Mission Anthropologique could have had a useful softening effect if it had successfully redirected Niger's medical staff away from coercive measures regarding childbirth and towards promoting nutritious indigenous plants and suitable weaning foods. It is hard to know whether the Mission researchers or reports had any direct impact in Niger – none of the medical reports for Niger mentioned the work of the Mission Palès in any fashion whatsoever. Nor is there any mention of passing researchers or of guidance on how to improve nutrition, or of efforts to integrate the many complex nutritional tables generated by ORANA into local practice. Given how stretched the health staff in Niger was, it is perhaps not surprising that it had little time to introduce nutrition projects into a health system that could barely meet the most urgent needs of the population. In order to reach women (whose role in food production would be central to any nutrition intervention), such efforts would almost certainly have had to be carried out through the existing structure of overstretched maternal health clinics, which, with only fourteen such centres in the entire country in 1952, would not, at any rate, have had much of an impact (Lorre 1952: 23). But as we have seen, very little of the Mission Palès' work was calibrated to address specifically maternal and infant nutrition in the first place.

By 1955 Niger did evidently have a service for the 'protection of children's health' but it was reported to be 'embryonic and barely functioning except in major centres' (Author unknown 1956). Annual medical reports for Niger, in the lead up to independence from the mid-1950s, were dismal affairs, dominated by uniform tables enumerating vaccinations delivered, consultations held, surgeries performed. More rarely, they included desultory commentary that simply repeated familiar conclusions: 'the principal causes of mortality among infants and children between

1 and 4 are the same as in the past, that is gastro-enteritis, malaria, respiratory infections, and smallpox' (ibid.).

Conclusion

In the records of the Service de Santé for Niger, one is left with the impression that the innovations in improving infant and maternal care that were to come to Senegal in the wake of the Mission Anthropologique, however limited and problematic they proved to be even there, entirely bypassed Niger. Medical officers in Niger were acutely aware of the problem of infant mortality and were attuned to the question of how food might be implicated in infant and child health, but they had neither the technical nor the budgetary means to begin to address it. It was only much later, with the creation of a dense network of centres for the promotion of maternal and infant health (PMI) during the Diori and especially the Kountché post-independence regimes, that any effective effort to reach women and children seems to have been made. The last of the files for Niger retained in the IMTSSA records a medical mission on the part of France to Niger in 1961. The occasion for this collaboration of a former colonial power and her newly independent protégée, sadly, was another meningitis epidemic. Among the factors that contributed to the crisis was 'chronic malnutrition complicated by multiple vitamin deficiencies particularly among children between two and six' (Lapeyssonnie 1961: 7).

If a variety of military style campaigns to improve the health of the population of Niger have been more or less successful in reducing infectious disease, the more difficult and longer term problem of coming to terms with the powerful driver of nutrition-immunity-infection interactions has yet to be fully appreciated. The impulse to see women's practices (from nursing to weaning) as the cause of children's malnutrition continues. Rather than contend with the political and economic contributors to poor health, such as the gaps in education for women, their declining access to farmland, inadequate health infrastructure in regions far from the cities, the poor water quality available in much of Niger, the hostility and expense of distant medical facilities, observers regularly point the finger at women's practices and 'culture' (Cooper 2007). The Mission Anthropologique researchers missed an opportunity to explore women's indigenous knowledge related to nutrition and farming, while administrators throughout the AOF gave very little attention to how changes prompted by the colonial economy might have affected food production, consumption and labour patterns. And in the end the rulers of this most marginal of AOF territories had little opportunity after the passage of the loi-cadre to draw upon the nutritional insights ORANA did uncover.

This study of the gendering of malnutrition in Niger offers an insight into a host of ways in which difference was configured in French colonial medical settings. While race was important, what race meant was quite ambiguous: at times it seemed to indicate innate physiological difference; at others it indexed ethno-linguistic or even more loosely cultural difference. Palès' disease mapping revealed that differences in susceptibility to disease were not linked to physiological difference; consequently, an emphasis upon cultural difference and loss of traditional knowledge came to the fore in place of a conception of race founded on biological difference. Nevertheless, over the long term the interest in racial difference linked to the larger concerns of physical anthropology and the demands of military recruitment delayed and deflected attention away from civilian women's needs, their bodies, and their situated expertise. The highly masculine quality of French colonial-cum-military medicine, with its preoccupation with male labour recruitment, rendered women's bodies and labour largely irrelevant. To the degree that women were seen as significant, it was as procreators – producers of black babies – or as culprits – mothers who fail to feed their families appropriately. They were not conceived as partners in promoting nutritional health through their skills in growing, transforming and cooking food.

Finally, if in France itself efforts to improve the social body were directed at the individual citizen, in the colonies populations were understood as 'races' (at first seen as biological and later as largely cultural) that were targeted for change. Research and policy generated in 'useful' territories and upon valued populations ultimately had very little significance in more marginal territories. The accumulated effects of different kinds of marginality were to leave women and children in Niger with the postcolonial legacy of a fragile medical infrastructure and a host of paternalistic assumptions about the cultural, and implicitly feminine, sources of poor nutrition and high infant mortality.

Barbara M. Cooper is a Professor and Chair of the Rutgers University Department of History (New Brunswick). Her major publications include *Marriage in Maradi: Gender and Culture in a Hausa Society in Niger, 1900-1989* (1997) and *Evangelical Christians in the Muslim Sahel* (2006), for which she was awarded the 2007 African Studies Association Melville J. Herskovits Award. Major grants from USDOE Fulbright-Hays, NEH and the Mellon Foundation have facilitated her research. She is currently completing a book manuscript tentatively entitled 'Countless Blessings: A History of Childbirth, Fertility and Population in the Sahel'.

Notes

1. This chapter is based largely upon documents accessed at the Institut de médecine tropicale du Service de santé des armées (antenne IRBA de Marseille, also known as the Pharo), where the most complete set of colonial health service reports for Niger are to be found. Other documents cited are to be found in the Archives nationales du Sénégal (ANS) which houses the colonial administrative records of the AOF. All translations are my own.
2. For a nuanced discussion of the history and philosophical groundings of neo-Lamarckianism, see Loison (2011: 713). On acclimatization theory, see Osborne (2000: 135). For French social hygiene and eugenics, see Schneider (1986: 265; 1990: 69; 2002).
3. For a glimpse into the exceptional quality of interventions in Cameroon, see Lachenal (2010: 121).

4. On the scope of the studies produced by the Mission, see Collignon (2006) and Collignon and Becker (1989).
5. Nutritional science has been built upon the study of research subjects whose diets could be bureaucratically controlled, notably sailors, prisoners, soldiers and orphans. See Carpenter (2003a: 638; 2003b: 975; 2003c: 3023).
6. See chapter 6 in this volume.
7. According to Adrian (1999), ORANA began with the physical and psycho-technical studies, turning only to nutritional issues as a second-phase concern.
8. On the shifting reception of this argument over several decades, see Muller-Wille (2010).
9. Researchers were struggling to understand how nutrition and disease were manifested in the liver, which complicated, as Bado notes in chapter 5, researchers' grasp of the incidence of liver cancer in African populations.
10. See J.P. Harroy (1949), *Afrique, terre qui meurt* (Brussels: M. Hayez), quoted in Palès (1954: 17).
11. By contrast Brock and Autret (1952) show growth curves for children in Kampala, where work on kwashiorkor in particular was given priority (14–15).
12. See chapter 8 in this volume.
13. Brantley (2002) found in the Nyasaland Nutrition Survey that a similar failure to bring together the 'measurable' data with relevant socio-political information meant that the study in the end had little practical impact.
14. Fonio is a variety of millet.
15. Ironically, prisoners also benefited from this attention, so that the prison population was claimed to be, if anything, better off than the average civilian population.
16. For an indispensable study of the training of African women to become *sage-femmes* [midwives] to begin to address the refusal of African women to attend clinics staffed by men, see Barthélémy (2010).
17. The annual inspection reports of the health services make reference to voluntary organizations across the AOF (Inspection General des Services Sanitaires de l'AOF 1932, 1934, 1940, 1948).
18. Occasionally, however, the medical staff did note that some populations were more receptive to medical interventions than others. Saliceti noted in 1941 that women in Agadez were quite comfortable with the idea of giving birth in a maternity clinic (74).

Bibliography

Adrian, J. 1999. 'La mission anthropologique et l'O.R.A.N.A', *Médecine et nutrition* 35(1): 18–21.

Appadurai, A. 1996. *Modernity at Large: Cultural Dimensions of Globalization*. Minneapolis: University of Minnesota Press.

Appel, Toby A. 1987. *The Cuvier-Geoffrey Debate: French Biology in the Decades before Darwin*. Oxford: Oxford University Press.

Autret, M. 1992. 'L'Œuvre du corps de santé colonial dans le domaine alimentation-nutrition Outre-mer', *Médecine et Nutrition* XXVII(5): 273–284.

Barona, J.L. 2010. *The Problem of Nutrition: Experimental Science, Public Health, and Economy in Europe, 1914-1945*. Brussels: P.I.E. Peter Lang.

Barthélémy, P. 2010. *Africaines et diplômées à l'époque colonial (1918-1957)*. Rennes: Presses Universitaires de Rennes.

Bonnecase, V. 2008. *Pauvreté au Sahel: La construction des savoirs sur les niveaux de vie au Burkina Faso, au Mail et au Niger (1945-1974).* Paris: Université de Paris.

——— 2009. 'Avoir Faim en Afrique Occidentale Française: Investigations et représentations coloniales (1920-1960)', *Revue d'Histoire des Sciences Humaines* 21(2): 151–174.

——— 2011. *Pauvreté au Sahel: du savoir colonial à la mesure international.* Paris: Karthala.

Brantley, C. 2002. *Feeding Families: African Realities and British Ideas of Nutrition and Development in Early Colonial Africa.* Portsmouth: Heinemann.

Brock, J.F. and M. Autret. 1952. *Kwashiorkor in Africa.* Geneva: World Health Organization.

Carpenter, K. 2003a. 'A Short History of Nutritional Science: Part 1 (1785-1885)', *Journal of Nutrition* 133: 638–645.

——— 2003b. 'A Short History of Nutritional Science: Part 2 (1885-1912)', *Journal of Nutrition* 133: 975–984.

——— 2003c. 'A Short History of Nutritional Science: Part 3 (1912-1944)', *Journal of Nutrition* 133: 3023–3032.

——— 2003d. 'A Short History of Nutritional Science: Part 4 (1945-1985)', *Journal of Nutrition* 133: 3331–3342.

Collignon, R. 2002. 'Pour une histoire de la psychiatrie coloniale française. A partir de l'exemple du Sénégal', *L'Autre* 3(3): 455–480.

——— 2006. 'Axe 1b: La mission anthropologique des populations de l'AOF: Etude des sources et place dans la science coloniale de l'après-guerre', *Terrain et archive*, 7 avril 2006. Retrieved 15 October 2011 from http://lodel.imageson.org/terrainarchive/document110.html

Collignon, R. and C. Becker. 1989. *Santé et population en Sénégambie, des origines à 1960.* Paris: INED.

Conklin, A. 1997. *A Mission to Civilize: The Republican Idea of Empire in France and West Africa, 1895-1930.* Stanford: Stanford University Press.

——— 2002. 'Faire Naître v. Faire du Noir: Race Regeneration in France and French West Africa, 1895-1940', in T. Chafer and A. Sackur (eds), *Promoting the Colonial Idea: Propaganda and Visions of Empire in France.* London: Palgrave, pp. 143–155.

——— 2013. *In the Museum of Man: Race, Anthropology, and Empire in France, 1850-1950.* Ithaca: Cornell University Press.

Cooper, B. 2007. 'La rhétorique de la "mauvaise mère"', in X. Crombé and J.-H. Jézéquel (eds), *Niger 2005: Une catastrophe si naturelle.* Paris: Karthala, pp. 199–228.

Crais, C. and P. Scully. 2009. *Sara Baartman and the Hottentot Venus: a Ghost Story and a Biography.* Princeton: Princeton University Press.

Cullather, N. 2007. 'The Foreign Policy of the Calorie', *The American Historical Review* 112(2): 337–364.

Debaene, V. 2014. *Far Afield: French Anthropology Between Science and Literature*, trans. J. Izzo. Chicago: University of Chicago Press.

Domergue-Cloarec, D. 1997. 'La prévention dans la politique sanitaire de l'AOF', in C. Becker et al. (eds), *AOF: Réalités et héritages, Tome 2.* Dakar: Direction des Archives du Sénégal, pp. 1228–1239.

Dozon, J.-P.. 2008. *Une anthropologie en mouvement: l'Afrique miroir du contemporain.* Versailles: Editions Quae.

Echenberg, M. 1991. *Colonial Conscripts: The 'Tirailleurs Sénégalais' in French West Africa, 1857-1960.* Portsmouth: Heinemann.

Ferembach, D. 1988. 'Leon Palès', *Bulletins et Mémoires de la Société d'Anthropologie de Paris* 5(XIV)4: 297–300.

Fogarty, R. and M.A. Osborne. 2003. 'Constructions and Functions of Race in French Military Medicine, 1830-1920', in S. Peabody and T. Stovall (eds), *The Color of Liberty: Histories of Race in France*. Durham: Duke University Press, pp. 206–236.

Gervais, R.R. and I. Mande. 2010. 'How to Count the Subjects of Empire? Steps toward an Imperial Demography in French West Africa before 1946', in K. Ittmann, D. Cordell and G. Maddox (eds), *The Demographics of Empire: The Colonial Order and the Creation of Knowledge*. Athens: Ohio, pp. 89–112.

Hardy, G. and C. Richet (eds). 1933. *L'Alimentation Indigène dans les Colonies Françaises, Protectorats et Territoires sous Mandat*. Paris: Vigot Frères.

Kamminga, H. and A. Cunningham. 1995. *The Science and Culture of Nutrition, 1840-1940*. Leiden: Brill.

Keusch, G.T. 2003. 'Symposium: Nutrition and Infection, Prologue and Progress Since 1968', *Journal of Nutrition* (Supplement)133: 336S–340S.

Labouret, H. 1933. 'L'Alimentation des Indigènes en Afrique Occidentale Française', in G. Hardy and C. Richet (eds), *L'Alimentation Indigène dans les Colonies Françaises, Protectorats et Territoires Sous Mandat*. Paris: Vigot Frères, pp. 138–154.

Lachenal, G. 2010. 'Le médecin qui voulut être roi: Médecine colonial et utopie au Cameroun', *Annales: Histoire, Sciences Sociales* 65(1): 121–156.

Lévi-Strauss, C. 1952. *Race and History*. Paris: UNESCO.

Loison, L. 2011. 'French Roots of French Neo-Lamarckisms, 1879-1985', *Journal of the History of Biology* 44: 713–744.

Lunn, J. 1999. '"Les Races Guerrières" Racial Preconceptions in the French Military about West African Soldiers during the First World War', *Journal of Contemporary History* 34(4): 517–536.

McCulloch, W.E. 1930. *An Inquiry into the Dietaries of the Hausas and Town Fulani of Northern Nigeria: With some Observations of the Effects on the National Health, with Recommendations Arising Therefrom*. Lagos: Government Printer.

Moore, H. and M. Vaughan. 1994. *Cutting down Trees: Gender, Nutrition, and Agricultural Change in the Northern Province of Zambia, 1890-1990*. Portsmouth: Heinemann.

Muller-Wille, S. 2010. 'Claude Levi-Strauss on Race, History and Genetics', *BioSocieties* (Special Issue BioHistories) 5(3): 330–347.

Ndao, M. 2005. 'Enfance et Ordre colonial. La politique sanitaire au Sénégal: discours et réalités (1930-1960)', *Revue Sénégalaise d'histoire* 5(6): 167–181.

——— 2008. 'Colonisation et politique de Santé Maternelle et Infantile au Sénégal (1905-1960)', *French Colonial History* 9:191–211.

Neill, D. 2009. 'Finding the "Ideal Diet": Nutrition, Culture and Dietary Practices in France and French Equatorial Africa, c. 1890s to 1920s', *Food and Foodways* 17(1): 1–28.

——— 2012. *Networks in Tropical Medicine: Internationalism, Colonialism, and the Rise of a Medical Specialty, 1890-1930*. Stanford: Stanford University Press.

Osborne, Michael A. 2000. 'Acclimatizing the World: A History of the Paradigmatic Colonial Science', *Osiris* 2(15): 135–151.

Palès, L. 1929. 'État actuel de la paléopathologie: Contribution à l'étude de la pathologie comparative', *Thèse pour le doctorat en médecine*. Bordeaux: Université de Bordeaux.

Palès, L. and M. Tassin de Saint-Péreuse. 1953. 'Raciologie comparative des populations de l'Afrique occidentale: stature, indice cormique, indice céphalique', *Bulletins et Mémoires de la Société d'anthropologie de Paris* X(4)3-4: 185–497.

Pépin, J. 2011. *The Origins of AIDS*. Cambridge: Cambridge University Press.

Schneider, W. 1986. 'Puericulture and the Style of French Eugenics', *History and Philosophy of the Life Sciences* 8(2): 265–277.

—— 1990. 'The Eugenics Movement in France, 1890-1940', in M. Adams (ed.), *The Wellborn Science: Eugenics in Germany, France, Brazil and Russia*. Oxford: Oxford University Press, pp. 69–109.

—— 2002. *Quality and Quantity: The Quest for Biological Regeneration in Twentieth-Century France*. Cambridge: Cambridge University Press.

Scott, J.C. 1999. *Seeing Like a State: How Certain Schemes to Improve the Human Condition Have Failed*. New Haven: Yale University Press.

Sibeud, E. 2002. *Une Science Imperial pour l'Afrique? La construction des savoirs africanistes en France 1878-1930*. Paris: Ecole des Hautes Etudes en Sciences Sociales.

—— 2012. 'A Useless Colonial Science? Practicing Anthropology in the French Colonial Empire, circa 1880-1960', *Current Anthropology* 53(5): S83–S94.

Simmons, D. 2008. 'Starvation Science: From Colonies to Metropole', in A. Nützenadel and F. Trentmann (eds), *Food and Globalization: Consumption, Markets and Politics in the Modern World*. New York: Berg, pp. 173–191.

Sorel, F. 1933. 'L'Alimentation des Indigènes en Afrique Occidentale Française', in G. Hardy and C. Richet (eds), *L'Alimentation Indigène dans les Colonies Françaises, Protectorats et Territoires Sous Mandat*. Paris: Vigot Frères, pp. 155–176.

Twagira, L.A. 2013. 'Women and Gender at the Office du Niger (Mali): Technology, Environment, and Food ca. 1900-1985', PhD dissertation. New Brunswick: Rutgers, the State University of New Jersey.

van Beusekom, M. 2002. *Negotiating Development: African Farmers and Colonial Experts at the Office du Niger, 1920-1960*. Portsmouth: Heinemann.

Worboys, M. 1987. 'The Discovery of Colonial Malnutrition between the Wars', in D. Arnold (ed.), *Imperial Medicine and Indigenous Societies*. Manchester: Manchester University Press, pp. 208–225.

Wylie, D. 2001. *Starving on a Full Stomach: Hunger and the Triumph of Cultural Racism in Modern South Africa*. Charlottesville: University of Virginia.

Archival sources

Institut de médecine tropicale du Service de santé des armées (antenne IRBA de Marseille) (Pharo). Box numbers follow the designation IMTSSA.

Author unknown. 1956. 'Rapport médical Année 1955, Territoire du Niger, 2eme partie' (IMTSSA 62).

Auffret, Le Pharmacien Commandant C. and le Pharmacien Commandant F. Tanguy. 1947. 'Vitamine A et Carotène dans le sang de femmes indigènes de la région de Dakar au moment de l'accouchement' in 'Rapport no. 3: Guinée Occidentale, Dakar, Sénégal et Soudan, AOF' (IMTSSA 398).

Bergouniou, Le Médecin Commandant J.L. 1951. 'AOF: Problème Alimentaire et nutritionnel' (IMTSSA 398).

Guillaume, Le Médecin Lieutenant-colonel. 1940. 'Rapport médical Année 1939, Colonie du Niger, Partie Administrative' (IMTSSA 62).

Kervingant, Le Médecin Lieutenant-colonel. 1948. 'Rapport médical Année 1947, Colonie du Niger, Partie médicale' (IMTSSA 62).

—— 1949. 'Rapport médical Année 1948, Colonie du Niger, Partie médicale' (IMTSSA 62).

Lapeyssonnie, Le Docteur L. Professeur agrégé des TOM, Chef de Mission Médicale française au Niger. 1961. 'Premier Rapport Magaria, le 10 Mars 1961' (IMTSSA 17).

Lorre, Le Médecin Lieutenant-colonel. 1951. 'Rapport médical Année 1950, Territoire du Niger, Partie médicale' (IMTSSA 62).

—— 1952. 'Rapport médical Année 1951, Partie administrative' (IMTSSA 63).

Mayer, A. 1954. 'Préface', in L. Palès, 'L'Alimentation en AOF' (IMTSSA 162).

Morvan, Le Médecin Lieutenant-colonel. 1946. 'Rapport médical Année 1945, Colonie du Niger, Partie médicale' (IMTSSA 62).

—— 1947. 'Rapport médical Année 1946, Colonie du Niger, Partie médicale' (IMTSSA 62).

Palès, Le Médecin Colonel L. 1945. 'Rapport préliminaire sur les travaux de l'Organisme d'Enquête Pour l'Etude Anthropologique des Populations Indigènes de l'AOF. (Alimentation-Nutrition)' (IMTSSA 162).

—— 1948. 'Le Bilan de la Mission Anthropologique de l'AOF 1946-1948' (IMTSSA 398).

—— 1954. 'L'Alimentation en AOF: Milieux, enquêtes, techniques, rations' (IMTSSA 162).

Palès, Le Médecin Colonel L. avec le concours de Mlle Tassin de Saint-Péreuse. 1949. 'Raciologie Comparative des Populations de l'AOF: I Parallèle anthropométrique succinct (stature) des militaires et des civils' (IMTSSA 398).

Pauliac, le Médecin Lieutenant-colonel. 1938. 'Epidémie de méningite cérébro spinale ayant sévi au Niger du 20 décembre 1937 au mois de mai 1938' (IMTSSA 17).

Saliceti, Le Médecin Lieutenant-colonel. 1941. 'Rapport Médical, Colonie du Niger, Année 1940' (IMTSSA 62).

—— 1943. 'Rapport médical Année 1942, Colonie du Niger' (IMTSSA 62).

Archives nationales du Senegal in Dakar (ANS) Documents of the AOF (Series H: Health). Series numbers are followed by the folder number and the 'versement' in which the set is to be found in Dakar.

Brévié, Gouverneur Général de l'AOF J. 1935. Memo to M. le Ministre des Colonies [Louis Rollin], 'Rapport sur la Protection du travail de l'enfance, le 28 juin 1935' (ANS 1H 102 versement 163 'Protection maternelle et infantile: organisation I').

Comité de la Journée des mères de familles nombreuses. 1920–1923. 'Correspondance' (ANS 2H 6 versement 26 'Journée des mères de familles nombreuses en AOF').

Inspection General des Services Sanitaires de l'AOF. 1932, 1934, 1940, 1948.

8

Medical Demography in Interwar Angola

Measuring and Negotiating Health, Reproduction and Difference

Samuël Coghe

In the 1920s, medical doctors in Angola, a Portuguese colony in West-Central Africa, began to collect and analyse demographic data on the colony's African populations in a far more systematic manner than before. Embedded in the context of new programmes of African healthcare, their studies used novel methods of data gathering and generated new insights into population dynamics that would play an important, though ambivalent, role in ongoing debates about the specificities of African population trends.

Medical demography, as I will call the demographic endeavours of colonial doctors here, was not confined to Portuguese Angola: it was a transnational phenomenon that can be observed in many colonies in Africa and beyond in the interwar years (and sometimes even a bit earlier). Yet the focus on Angola, Portugal's largest colony and for many observers in the nineteenth and twentieth centuries the cornerstone of Portugal's Third Empire due to its long colonial history and its economic potential (Valentim 2000; Santos 1945), allows for a particularly insightful case study. It illuminates why and how medical demography emerged and developed as a transnational practice that circulated between colonies and empires and how demographic intelligence contributed to debates about reproduction and difference.

In itself, the engagement of doctors with demography was not a colonial specificity. Before demography was institutionalized in Europe and North America as a discipline in its own right around the Second World War, i.e. when proper research institutes, journals and professional career paths were established (Greenalgh 1996: esp. 30, 34–41; Rosental 2007), medical doctors were often at the forefront of demographic research. Internationally, the International Congresses on Hygiene and

Demography (1878–1912) and the collection and publication of demographic data and analyses by the League of Nations Health Organization (LNHO) epitomized the close, though not exclusive, connection that existed between the two fields since the late nineteenth century (Rosental 2007: 257; Overath 2011: 66–68; Borowy 2009: 177–178). In Portugal, it was the Central Institute of Hygiene in Lisbon that was in charge of publishing and analysing the country's official demographic data in the 1910s and 1920s (Sousa 1995: 161–163). Its director, Ricardo Jorge, was not only one of the country's most recognized hygienists, but also one of its leading experts in demography, just like his colleague António de Almeida Garrett in Porto (Baptista 2007: 540–541; Ricon-Ferraz and Guimarães 2008: 9–14).

Yet the demographic work of colonial doctors in interwar Africa differed from that of their colleagues in Europe in various regards. Whereas doctors in Europe could base their studies on reasonably accurate pre-existing data from censuses or civil registries, doctors in colonial Africa usually still had to collect themselves the data they wanted to analyse, especially in rural areas. Counting people in the very literal sense became an important part of their job. In the 'colonial situation', doctors were thus turned into 'field demographers', as opposed to the 'armchair demographers' up in Europe.[1] This also had implications for the way in which they analysed their data. Since many of their demographic reports were primarily written for practical purposes of medical control and policy making, and less for academic purposes, they did not always explicitly and intellectually engage with the issue of racial difference, which was central to debates among academics in Europe studying the demography of colonial populations (Ittmann 1999: 57–58, 64–65; Widmer 2014: 547–548).

That does not mean, however, that racial thinking was absent from their writings. On the contrary, colonial doctors' demographic work in Angola both reflected and contributed to ideas and debates about racial difference in various ways. Assumptions about differences in fertility and mortality patterns between Africans and Europeans fundamentally underwrote medical demography, since it emerged and was mostly practiced within the context of healthcare programmes that were exclusively directed to the *indígenas* ('natives'), that is those Africans who were still considered 'non-civilized'.[2] Their presumed demographic decline, moreover, constituted the main rationale for both the healthcare programmes and the demographic studies. Tellingly, doctors' research into the demography of the colony's 'native' population was not paralleled by similar efforts towards studying the demographic behaviour of European settlers and/or racially mixed communities in Angola before the 1940s.[3] Demographic knowledge production was largely confined to the *indígena*, the 'non-white' and 'non-civilized' Other, whose difference was assumed on the basis of both biological and cultural traits.

Conversely, medical demography also influenced racial thinking. It contributed to the refashioning and entrenching of ideas about the different demographic regimes that existed between the *indígenas* in Angola and people in Europe on the one hand, and (to a lesser extent) also in-between the colony's *indígenas* on the other hand, as colonial doctors compared their data with those existing on other populations or analysed them through the lens of presumably universal standards. Hence, even when ideas about difference were not always visible or fleshed out, interwar medical demography was clearly part of a broader colonial effort towards studying and categorizing the different populations of the Portuguese empire, an effort to which other disciplines such as physical anthropology were also – and more explicitly – contributing (Roque 2001, 2010; Matos 2006; Santos 2012).

While this article speaks to this literature on colonial difference and governance, it also contributes more generally to the existing historiography on colonial demography. Firstly and most obviously, this historiography has completely overlooked the existence and importance of medical demography in colonial Africa. This applies to both African historical demography, a discipline that uses colonial censuses and other demographic data primarily as sources with which to reconstruct the demographic and socio-economic past of societies, and a new historiography that, from the mid-1980s onwards and under the influence of the cultural and postcolonial turns, has challenged the positivism of historical demography.[4] Both strands of research have mainly focused on administrative censuses, with historical demography also using demographic data gathered by missionaries and travellers.[5]

Secondly, cultural and postcolonial approaches have certainly added new perspectives to the history of demography. Studies have analysed the colonial census as an instrument of knowledge for the colonial state; explored the complexities of colonial census-taking on the ground; highlighted the importance of ethnic and social categories in the census for the classification and rule of populations by colonial states; argued that the flood of numbers was also used to create and maintain the illusion of bureaucratic control; and, in some cases, pointed at the continuities between pre-colonial, colonial and post-colonial census-taking methods (Cohn 1987; Appadurai 1993; Peabody 2001; Gervais and Mandé 2007). The question of how colonial demographers themselves viewed, interpreted and used their data, however, has received very little attention. Hence, by addressing these questions with regard to medical doctors in interwar Angola, this chapter is also a genuine contribution to the cultural history of colonial demography.

In a first step, I show how the emergence of medical demography in Angola in the 1920s was conditioned by innovations in African healthcare as well as persistent anxieties of population decline and facilitated by the inter-imperial transfer of demographic methods already in use in other colonies. I then demonstrate that, despite the often well-known shortcomings of these methods, doctors did not refrain from producing demographic analyses that served their own purposes and that had a significant impact on population debates in Angola. Not only did doctors suggest that the ongoing investments in healthcare had stopped population decline, they also claimed that, due to civilizational backwardness, Africans were governed by a demographic regime of high fertility and high infant mortality that was distinct from the one in Europe.

The Emergence of Medical Demography

In Angola, the rise of medical demography was inextricably linked to the establishment of the *Assistência Médica aos Indígenas* (AMI) ('Native Medical Assistance'). In 1926, after years of discussion, the colonial government in Angola had created this scheme of rural African healthcare in the northern districts of the colony (Zaire, Congo, Cuanza-Norte), where sleeping sickness was raging and depleting the populations. Largely inspired by healthcare programmes in other, notably French and Belgian, colonies in Africa, the AMI programme was the most systematic colonial healthcare effort in rural Angola thus far. It integrated the fight against sleeping sickness, a deadly tropical disease that had drawn continuous attention from all colonial powers in Africa since the turn of the twentieth century,[6] into a more comprehensive scheme that, at least in theory, aimed at addressing all major diseases and health problems weighing on the rural African populations. Therefore, the northern districts were divided in four zones and further subdivided in twelve, later fifteen sectors, in which a gradually expanding number of health centres and village-infirmaries were established. Each of the sectors was directed by a medical doctor who, assisted by European and African nurses, was bound to tour around and regularly examine and treat its entire population (see Coghe 2015).

The link between medical demography and the AMI programme was twofold. Profoundly influenced by ideas of collective and preventive hygiene, often subsumed under the term of social medicine, leading proponents of the AMI programme, like the director of the health services António Damas Mora, emphasized the importance of demography. In their view, accurate data on demographic trends was key for identifying health problems, for adapting the AMI programme to the needs of the population and for monitoring its impact (Santos 1923: 52–54, 70; Costa 1928b: esp. 104; Mora 1934b: 39–40). In the words of João Camoesas, a prominent Portuguese hygienist employed by the AMI services in the late 1920s, the survey of the population was 'the fundamental basis of all socio-medical action' (1929: 143). Conversely, the AMI programme also offered unprecedented opportunities for demographic studies, as it brought doctors into more regular and hierarchically structured contact with rural African populations.

The emergence and spread of medical demography in interwar Angola must also be understood against the backdrop of the widespread idea of a demographic crisis. Since the beginning of the twentieth century, anxieties of depopulation pervaded European colonial discourse with regard to most colonies in tropical Africa and the Pacific (see Coghe and Widmer 2016: 37–64). Many doctors, administrators and missionaries believed that the 'native' populations in these colonies were declining. They blamed a wide array of causes: epidemic and endemic diseases, exploitative colonial labour regimes, general ignorance and a lack of hygiene among the 'native' populations and even deliberate 'racial suicide'. These conditions, they assumed, were responsible for a 'native' demographic regime characterized by low birth and high mortality, especially infant mortality, rates (Athayde 1918; Sant'Anna 1923: esp. 73–74; Matos 1926: 227–277; see also Legrand 1921; Peiper 1920: 4). However, even in the early 1920s, when in many parts of Africa fears of depopulation resurged due to the economic and political imperatives that the colonial ideology of *mise en valeur* implied (Headrick 1994: 104–105), reliable data on the mortality and natality of African populations living outside the few larger towns were still scarce or non-existent. Hence, besides triggering more comprehensive native healthcare programmes that would halt and reverse the presumed population decline in the first place, depopulation fears also turned the production of more reliable demographic data into an urgency.

This does not mean that there was a complete dearth of demographic data on Angola's 'native' population. There was a long tradition of administrative census taking in colonial Angola, which reached back to the 1770s, when Marquês de Pombal, the 'enlightened' Prime Minister of Portugal, ordered the first surveys of the population in Portuguese-ruled Angola. For decades, censuses were carried out almost every year in the colony's coastal regions, but with the end of Angola's official participation in the transatlantic slave trade in the 1840s, interest in demography waned (Curto and Gervais 2001).[7] This only changed again at the turn of the twentieth century, when the scramble for Africa and the gradually emerging international imperative of scientific colonialism spurred the collection of demographic data. In 1899, an ambitious law for decennial censuses throughout the Portuguese Empire was adopted (Ministério da Marinha e Ultramar 1900: 343).[8] Certainly, the problems encountered during the census operations of 1900, such as the still fragmentary control over the colony's territory, the lack of personnel and the resistance of African subjects, proved insurmountable and the dream of a colony-wide autonomous scientific census would only be fulfilled in 1940 (Governo Geral da Província de Angola, Secretaria Geral 1903: 161; Repartição de Estatística Geral da Colónia de Angola 1941–1947). But as the effective occupation of the interior of the colony advanced, more and more local administrators produced annual population statistics, which were based on tax registers and, from 1913 onwards, collated in colony-wide 'census' maps by the newly established Native Affairs Department. From the mid-1920s onwards, these administrative censuses covered the entire colony (see Diniz 1914: 11–16, 1930; Heywood and Thornton 1987). In addition, the Angolan government conducted several surveys in the 1920s and 1930s to determine the number of male adults in working age (Matos 1923: 314–316; Carvalho 1925: 46; Ferreira 1927: 49–53).

From the vantage point of medical doctors, the problem, hence, was not so much the lack of demographic data, but its quality. In the absence of autonomous censuses and a functioning civil registry – registration would only become obligatory for the African population in Angola in 1942 – doctors in Angola, just like their colleagues in many other African colonies, neither trusted the available data nor considered them useful for their purposes.[9]

On the one hand, they complained that administrative censuses were erroneous and of little scientific value, because of various methodological and practical problems and above all the very rationale that underpinned them: i.e. the aim to measure the healthy adult male population for the sake of tax collection, labour recruitment and military conscription. Doctors claimed that many adult men tried to hide from the administrator to avoid registration and its negative consequences, and that administrators often deliberately overestimated the age of male adolescents so that they could declare them as recruitable and tax-paying adults. According to doctors, the focus on adult males also led to the undercounting or underestimation of the numbers of women and children. They accused administrators of not ensuring that all women and children were actually counted and of just deducing the number of women and children by making a mathematical calculation based on the number of adult men, which, moreover, they deemed to be generally too low (Camoesas 1929: 143–144; Silva 1936: 88–89, 92; Mora 1934b: 39).[10] On the basis of these criticisms, which are strikingly similar to those formulated by some Africanist historical demographers decades later (see Heywood and Thornton 1987: 243–244, 250–251), doctors legitimized their own censuses and explained why these came up with different numbers. This was, for instance, the case of Venâncio da Silva, who had found fewer adult men, but 30 to 50 per cent more women and children during his medical census in two administrative posts in the Congo Zone in 1928, in comparison with the data of the administrator (Anonymous 1929a: 298–299).

On the other hand, doctors also criticized administrative censuses for only providing a static view of the population. As hygienists, they were more interested in population dynamics than in total numbers; they wanted to know whether populations were growing or declining, now and in the future, and to understand why that happened. For this purpose, they needed death, birth and migration rates and more specific indices such as infant mortality rates, ratios of births per woman, sex ratios, age group distributions as well as morbidity statistics indicating the incidence of certain diseases. Administrative reports and population maps did not generally contain such information, nor data that was sufficiently secure for doctors to calculate these indices themselves (Camoesas 1929: 143–144; Mora 1934b: 39).

For these reasons, doctors began to use the framework of the new AMI programme to raise their own data, thus becoming 'field demographers'. If, conversely, doctors did not engage in similar efforts with regard to Angola's white European population, that was not only due to the absence of a similar depopulation discourse, but also largely to the absence of a similarly comprehensive and authoritarian healthcare programme that would have brought them in regular contact with Europeans and given them leverage over them. Europeans were either treated on an individual and voluntary basis or often, when they lived far from the next doctor or could not afford medical treatment (which was except for civil servants not free to them), not at all (Dias 1933: 2).

The Methods of Medical Demography

AMI doctors in Angola basically used two different methods to measure the African population and their reproductive behaviour: a registration system and oral interviews.

In the AMI sectors of northern Angola, where the health services pursued the utopian vision of complete and continuous medical control, the aim was nothing less than to register the entire population and its 'movements': births, deaths and in- and outbound migration. To do so, the doctor and nurses in charge of a sector made use of the system of 'concentrations' (*concentrações*) that had been established to combat sleeping sickness (and other diseases). At fixed moments in time – usually every two weeks or once a month – all people from a certain area were required to appear at a predefined place in walking distance from their villages, under the threat of otherwise being punished with forced labour. At these concentrations, they were not only examined, treated and if necessary sent to local infirmaries, they were also registered. Registration was compulsory, nominative and double: each person was registered on a card (*ficha sanitária*), which he or she was supposed to keep, and also in the registry books, which were kept by the medical authorities of the sector (see Anonymous 1929a: 296–297). Doctors also recorded the deaths and migrations that were reported to them. Particular attention was paid to the registration of births. In order to keep track of newborns, doctors not only registered the births that were reported to them, but also took note of ostensibly pregnant women during the concentrations. They wrote down the probable moment of childbirth and exhorted expectant mothers to bring their babies to them once they had been born (Teixeira 1931: 8, 15; Ornelas and Mesquita 1935: 24).

This registration system, which constituted a kind of medical counterpart to the inexistent civil registry, was initiated in the Cuanza Zone in 1927 and gradually extended to all zones and sectors in the following years. By the end of 1930, the AMI services had registered almost 400,000 people, a number that would rise to almost 600,000 by the end of the decade (Teixeira 1931: 16, 19; Silva 1930: 45; Direcção dos Serviços de Saúde e Higiene da Colónia de Angola 1941: 63). Although this registration scheme would provide the raw material for many a demographic study, its aim was not merely demographic. Along with demographic data, the *fichas* and books also contained information on diseases and treatments. Some doctors, like João Camoesas, even wanted to further enhance the potential of this 'medical registry'. Camoesas, who had previously worked as a hygienist in Lisbon and been deported to Angola for political reasons in 1928, recommended that additional medical data as well as basic physiological information, ranging from height and weight to the much en vogue Pignet index – an index that evaluated physical robustness by relating thorax circumference to height and weight – be registered. To facilitate the identification of individuals, he also proposed to introduce metal bracelets similar to those worn by soldiers during the First World War. If well organized, he claimed, such a medical registry would generate a huge amount of scientific data to study the vitality of the population, while simultaneously facilitating tax collection and rationalizing labour recruitment (Camoesas 1929).

Camoesas' ideas did not entirely go unheeded. In 1930, the head of the Cuanza Zone, Waldemar Teixeira, adopted Camoesas' model of register cards with minor changes for the area around Cazengo. Accordingly, he measured more than 5,000 people from both sexes and all ages and calculated their Pignet index. In a few hundred cases, he also determined the so-called Lefrou index, which had just been developed by the French colonial doctor Gustave Lefrou in order to adapt Pignet's Eurocentric formula to the 'particularities' of the African body. Yet, even according to this index, most adult males were considered too frail to be recruited (Teixeira 1931: 27–32;

Lefrou 1931). Even if this time-consuming procedure was not extended to all AMI sectors, Camoesas' recommendations show to what extent hygienists in the interwar period dreamt of a 'legible' population for the sake of medical and administrative governance.

In order to get a better grasp of the long-term biological reproduction levels of African populations, some doctors also resorted to another method. During the concentrations, they questioned women in and beyond childbearing age about their reproductive life. In these oral interviews, which were mostly conducted with the help of an interpreter, doctors asked how many children they had born, how many of them had died in early infancy or later, and how often they had aborted or born stillbirths. With this information, they calculated the total and age-specific number of children per woman, determined the level of infant mortality and mapped the incidence of abortions and stillbirths.

In contrast to the medical registry, interview-based fertility and infant mortality studies were sample studies. They encompassed small groups of a few hundred to a few thousand women who were considered representative for a particular region or ethnic group in the colony. They were conducted in the AMI sectors, where they added historical depth to the registry data, but also during mobile medical missions in regions that were not yet under systematic medical control (Teixeira 1931: 15–26; Amaral 1939: 172; Rezende 1929).[11] In these latter areas, the interviews were the only means with which to obtain an approximate idea of fertility and infant mortality levels, and hence of population trends. Thus, even the two doctors who accompanied the Roads Study Mission (*Brigada das Estradas*) that was dispatched to Southern Angola in the early 1930s seized the opportunity to proceed to several studies of this kind. Both had previously worked in AMI sectors (Freitas and Fonseca 1930: 8–12; Fonseca 1931: 2–11; Freitas 1931: 5–11; 1932: 2–5). That interview-based fertility and infant mortality studies became a common feature of medical demography demonstrates the great importance the Angolan health services attached to these issues, but also the relative simplicity of the method.

It is important to note that neither the idea of a medical registry nor the technique of small-scale fertility and infant mortality studies originated among Portuguese doctors in Angola. Haunted by similar fears of low birth rates and high infant mortality, doctors in other African colonies had already introduced these novel forms of demographic inquiries. As early as 1909, German doctors began to conduct interview-based fertility and infant mortality studies among the African populations in Kamerun, Togo and German East Africa (Külz 1910: 33–35; Rodenwaldt 1915; Peiper 1920). After the First World War, Belgian, French and British doctors followed their example (Schwetz 1923: 325–340; 1924; Mouchet 1926; Clapier 1920; Bauvallet 1931; Cazanove 1930; Kuczynski 1939: 272, 282–286, 519–520). Moreover, this technique was also used in other continents, as examples from French Indochina show (Chesneau 1937).

While fertility inquiries could be conducted virtually everywhere, medical population registries required a much higher level of control. In rural Africa, the medical sectors that were established to combat sleeping sickness, and later other endemic diseases too, constituted a particularly propitious, though not exclusive, environment. Such sectors were first established in French Equatorial Africa (AEF) in 1917 and subsequently in French Cameroon (Piot 1920; Jojot 1921; Headrick 1994: 345–356), but it seems that they were first used for an encompassing population registry when the sector system was adopted in the Belgian Congo by Jacques Schwetz in the early 1920s (Schwetz 1923: esp. 299–300). It was also in the Belgian Congo that the FORÉAMI, a sector-based programme of African healthcare established in the Bas-Congo region in 1931 and similar to the AMI scheme in Angola, would use the technique of medical registries to elaborate what were probably the most extensive and detailed studies of medical demography in interwar Africa (Trolli 1934; Trolli and Dupuy 1934).[12]

Clearly, demographic techniques circulated across colonial and imperial boundaries and there is no doubt that inter-imperial learning was crucial for their introduction in interwar Angola. As I have shown elsewhere, the health services in Angola, and most notably its director António Damas Mora, were eager to emulate best practices in healthcare and hygiene from other colonies (Coghe 2015). This included demographic techniques: during a study tour in West Africa in 1926 organized by the League of Nations Health Organization (LNHO), Damas Mora had attended a pioneering lecture on medical demography in French-mandated Togo (see Mercier 1926). And when he later wrote a lengthy report on Native Medical Assistance in tropical Africa for the same LNHO in 1929–1930, he encountered many further studies in medical demography on French, British and Belgian colonies.[13] Moreover, many of the demographic studies by doctors in Angola openly referred to similar studies in foreign colonies as examples and role models (Ornelas and Mesquita 1935: 23–24; Teixeira 1931: 13–14; Fonseca 1931: 9). Inter-imperial learning thus filled a double gap: during the first decades of the twentieth century, Portuguese doctors had generally received little to no training in demography before going to the colonies, as it was not included in the curriculum of Portugal's medical faculties nor in that of the School of Tropical Medicine in Lisbon.[14] Only in 1960 would a first comprehensive manual on statistics and demography for Portuguese colonial doctors be published (Reis and Sarmento 1960). Moreover, doctors in Angola could not benefit from inner-imperial transfers, since field studies in medical demography only appeared in other Portuguese colonies later on (see, for instance, Ribeiro 1933: 3–14).

The Tensions of Medical Demography

Medical demography did not automatically lead to a more accurate view on population trends. First of all, the practical implementation of medical registries and retrospective interviews in rural Angola challenged the boundaries of European medical power; its success depended on several intersecting negotiation processes between AMI personnel and local actors. Furthermore, the interpretations doctors made of the collected data were often void of scientific rigour and were even tendentious, as they were used to serve a particular purpose.

The most basic problem was that, for both methods, the accuracy of the demographic data was fundamentally contingent upon the collaboration of the African population. Certainly, some doctors displayed much confidence in the reporting strategies of the Africans in their medical sector. Alfredo Gomes da Costa, the first director of the Cuanza Zone, for instance, asserted that probably only 2 per cent of the population in the entire zone was not in regular contact with the medical personnel (Costa 1928b: 104). Similarly, the young Goan doctor Eduardo Diniz da Gama claimed that, in his sector, only very few persons failed to attend the concentrations and that expectant mothers readily reported their pregnancy to the AMI personnel. These cases, he concluded, were duly registered and 'all pregnant women thus become responsible towards us for the foetus and, in the case of abortion, they come and tell us. On this point, some even say jokingly that the census of the [medical] assistance [services] is so complete that the blacks [*pretos*] are even registered when they are still in the belly of the mother' (Gama 1929: 541). Many other doctors, however, were far more sceptical and expressed their frustration over the reporting strategies of 'their' Africans. While Africans reported deaths with relative ease 'because of the benefits associated with such a declaration', that is the exemption of tax payment, they were much more reluctant to declare births, out of fear for future tax payments, a common complaint went (Sousa n.d.: 16; Teixeira 1931: 15; Mora 1940: 575–576).[15]

Even if their assessment of the current attendance and reporting rates of the Africans in the sectors diverged, doctors were unanimous in their analysis of how to (further) improve these rates, and hence the accuracy of their demographic data. They agreed that Africans' distrust towards medical demography was the main cause of avoidance and underreporting tactics and that, consequently, the cultivation of trust towards their actions was the key to success. It was imperative, they stated, to convince the Africans that the registration was an autonomous operation for medical and demographic purposes only and that, in other words, the data would not be used for the purposes of taxation, labour recruitment or conscription (Sousa n.d.: 16; Mora 1940: 575).[16] Trust was also important in another sense. Because of the simultaneity of registration and medical treatment, the quality of the demographic data also depended greatly on the trust Western biomedicine inspired among the African population.

The success of medical demography, and of the AMI programme in general, not only hinged on the persuasive power of the Portuguese doctors and nurses, but also on the collaboration of intermediaries, such as African nurses and local chiefs, as well as missionaries and local administrators. These groups played a crucial role in mediating power and trust between Portuguese doctors and the African population (Gama 1928: 60–63). Effective collaboration, however, was far from given. 'Native' nurses, for instance, were often praised by AMI doctors for their services as interpreters or for even actively convincing Africans of the advantages of Western biomedicine (Anonymous 1929b), but there were also frequent complaints about them following their own agenda and sometimes even abusing local populations, if they were given too much power and autonomy (Costa 1928a: 34).[17] Local administrators, then, were bound to ensure that all people effectively appeared at the concentrations and to assist the AMI doctors in other matters too, but conflicts between doctors and administrators frequently thwarted efficient collaboration (Costa 1928b: 104; Sousa 1928: 38, 68-69). Where doctors could not rely on local alliances to dissipate mistrust and, if necessary, to punish no-shows, the concentrations often remained empty (Freitas and Fonseca 1930: 6–7; Freitas 1931: 2, 8).

Finally, compared to the medical registries, retrospective interviews with women allowed for even fewer possibilities to effectively control the numbers. Here, intelligence was fully dependent upon the informants – upon their understanding of the questions and upon their memories, but also upon their goodwill in not under- or over-reporting their childbearing practices (Teixeira 1931: 24; Sousa n.d.: 16). Women had various motives to retroactively reshape their fertility, ranging from mistrust towards colonial intrusions to the unwillingness to confess childlessness in front of other women, given the high prestige that was assigned to motherhood and the grave dishonour associated with childlessness (Mora 1940: 588; similarly, Cazanove 1930: 36).

The contingencies of medical demography created a dilemma for the AMI doctors involved. While most of them were, at least to some extent, aware that the demographic data they had themselves collected or received from subordinate doctors could not be entirely accurate, they also knew that it was almost impossible to overcome the biases in the production of such and that there was no better data available. Unless they wished to discard the very possibility of studying population dynamics in rural Angola, they had little other choice than to use their data.[18] Indeed, most doctors in Angola solved this dilemma by eventually ignoring or downplaying the inaccuracy of their data. When it came to interpreting it, they presented the numbers, percentages and indices as positivist truths and often even based bold claims on them.

This phenomenon was not confined to Portuguese doctors in Angola. It was a basic tension of medical demography in other colonies as well, and of colonial demography in general. Already in 1937, Robert René Kuczynski, a prominent German demographer of Jewish origin who had started working on colonial populations after his emigration to London, criticized the 'appalling' extent to which 'the authors of colonial reports are tempted to draw far-reaching conclusions from the scanty population data at their disposal' (Kuczynski 1937: xii). In his detailed studies on demography in British, French and former German colonies, Kuczynski was very critical of most kinds of demographic methods that had thus far been used in colonial Africa. He also repeatedly pointed at the methodological shortcomings of the fertility and infant mortality studies conducted by colonial doctors (Kuczynski 1939: xiv, 42–43, 138–142). Yet, for Kuczynski, the problem was not only how data was gathered; he also criticized the fact that many colonial officials mis- or over-interpreted demographic indices because of their lack of basic skills in statistics or, more or less willingly, in order to support their views (Kuczynski 1937: xii–xiii).

The interpretation of the birth and death rates registered in the AMI Cuanza Zone is an illuminating example of this. In his annual report for 1930, the head of the Cuanza Zone, Dr Waldemar Teixeira, drew attention to the fact that the registered birth and death rates had both increased between 1927 and 1930, from 8.4 to 40.81 per thousand and 11.4 to 29.58 per thousand respectively. While he admitted that this increase was mainly the consequence of a better registration quote, he also interpreted the figures as proof of a distinct demographic shift. Under the impact of the AMI health programme, he claimed, population decline, still visible in the negative growth rates of 1927, had slowed in 1928 and reversed into population growth in 1929, a trend that was further consolidated in 1930 (Teixeira 1931: 14–15, 18–19).

In Teixeira's opinion, two other indices corroborated his claim. First, there was the shifting distribution of the population by age: between 1927 and 1930, he wrote, the percentage of people aged 0 to 15 had increased from 35.45 to 43.57 per cent, and thus changed the prospect for the future evolution of the population from 'stationary' to 'clearly progressive' (ibid.: 23).[19] And second, during the same time span, the sex ratio for the total population had fallen from 110 (men to 100 women) to 84. Not only had the sex ratios in the Cuanza zone thus approached global standards, they had also turned positive (ibid.: 22). Indeed, many a colonial doctor and demographer interpreted a low sex ratio (that is, more women than men) as an indication of future population growth, arguing that women were more important for reproduction than men and that the opposite, a high sex ratio, had been proven to be a symptom of racial degeneration and population decrease, like in some Pacific islands (Sant'Anna 1931: 19–21, 27; Mora 1940: 574).[20]

Although he knew that the data for 1927 and 1928 was very fragmentary, Teixeira did not admit the possibility, at least not overtly, that proportions were distorted or that the population had already been growing before the establishment of the AMI programme. His positivist analysis of the scanty data available was not an isolated case. A year earlier, his predecessor Augusto Ornelas had already analysed the birth and death rates in the Cuanza zone in very similar terms (Ornelas and Mesquita 1935: 22–23).[21] And this alleged reversal from population decline to modest, but continuous population growth, which would be confirmed by the medical registries in the Cuanza zone throughout the 1930s (Costa 1935: 61–62; Direcção dos Serviços de Saúde e Higiene da Colónia de Angola 1941: 65), also figured prominently in the writings of the health director, António Damas Mora (Mora 1934c; 1940: 216, 221). There was a clear rationale to their claims: by giving credence and publicity to this data and its interpretation in both internal reports and press articles, doctors aimed to legitimize the health investments and the particular methods of the AMI programme to the general government of the colony and the wider public.

The impact of these claims on the depopulation debate is more difficult to measure. For sure, they did not dissipate depopulation fears in Angola wholesale. The AMI zones had often been considered the worst regions in term of demographic evolution, but they only covered a small part of the colony. And depopulation anxieties were constantly reinforced by the increasingly pervasive idea that numerous Africans were emigrating to neighbouring colonies (Coghe 2014: 381–442). Yet when the director of the colony's newly established statistical services, Alberto de Lemos, published new population estimates for the whole of Angola in the mid-1930s, his figures both followed and reinforced the AMI claims, as they suggested a continuous population decline from the mid-nineteenth century until exactly 1927. From then onwards, Lemos explained, the population had begun to grow mainly because of the AMI programme (Repartição Central de Estatística Geral de Angola 1936: 21–23). Unsurprisingly, Damas Mora publicly endorsed this view later on (Mora 1940: 579–583).

Measuring and Debating Fertility, Infant Mortality and Difference

Medical demography also reshaped the debate on fertility and infant mortality in interwar Angola. Up to that point, both low fertility and high infant mortality had frequently been mentioned as major causes of population decline (Athayde 1918: esp. 230, 237; Sant'Anna 1923: 74; Matos 1926: 247). Now, on the basis of the birth rates and the average number of children per woman calculated in medical sectors or during mobile missions, many a doctor concluded that, in most parts of the colony, fertility was high, even much higher than in the metropolis (Mora 1940: 595; Sarmento 1944: 2), and therefore could not be in itself a cause of depopulation. Antero Antunes do Amaral, for instance, had personally registered birth rates between 44 and 58 per thousand and total rates of seven pregnancies per woman in the different sectors of the Cuanza Zone where he worked during the 1930s. For him, the matter was clear: 'There is in fact no natality problem to resolve', he stated emphatically (Amaral 1939: 172).

In the memoir he presented at the Colonial Congress in Lisbon in 1940, António Damas Mora basically argued along the same lines, but he presented a more complex picture. Inquiries had shown, he wrote, that women above the age of fifty had, on average, born five to six children during their reproductive life, 'with the exception', however, 'of regions where the native suffers from pronounced and permanent undernutrition' (Mora 1940: 573–574). The example he gave for this were the so-called 'Bushmen' in South Angola, often also designated as Khoisan or !Kung at the time. Due to the 'permanent state of undernutrition' and 'extreme misery' in which they lived, the Bushmen suffered from exceptionally low birth rates and were on the brink of extinction, Damas Mora stated (ibid.: 595–596). To this socio-medical explanation for differential fertility, Damas Mora also added racial factors. Endorsing the widespread idea, first formulated by anthropologists in the late nineteenth century, that the Bushmen were the vanishing remnants of an older race that had been repelled into inhospitable regions by the territorial expansion of Bantu peoples and Boers (see Bertillon 1882: 1; Diniz 1918: 479–491), he assumed that 'low fertility can also mean the senescence or ageing of a race'. Maybe, he concluded, the demographic decline of the Bushmen was, similar to that of the Australian aborigines, the result of their 'collective old age' and they would disappear just as the Tasmanians had in the nineteenth century. The natives of the 'Bantu lineage', the overwhelming majority in Angola, by contrast, had kept the 'psychology of the primitive peoples concerning the advantage of having many children' (Mora 1940: 595–596).

With this last remark, Damas Mora pitted the reproductive behaviour of Angola's Bantu populations against the decline of fertility that had been diagnosed for most European countries in the late nineteenth and early twentieth century and had provoked widespread anxieties of degeneration and depopulation, even in Portugal (Tomlinson 1985; Soloway 1990; Costa-Sacadura 1923). But he may also have been referring to the theory of deliberate racial suicide, which W.H.R. Rivers had advanced two decades earlier to explain the population decline in Melanesia. Rivers had posited that Melanesians, with the exception of those who had become Christians, refused to have children because they had lost interest in life due to the destruction of their traditional customs and the new hardships brought about by colonial rule. To avoid procreation, they made wide use of contraceptive and abortive practices (Rivers 1922; Widmer 2012). Echoing much older anxieties over the use of abortifacients by African slave women in the eighteenth- and nineteenth-century Atlantic world (Schiebinger 2004), this psychological explanation of population decline became an influential theory in the interwar period. Nancy Rose Hunt has shown that, due to the '"global circulation" of such "ethnographic commonplaces of colonial intelligence"', tropes of 'racial suicide, dying races, empty villages and self-aborting women' also widely circulated among French and Belgian colonial officials in Central Africa (Hunt 2007: 252–254, quotes 254).

In Angola, stories about abortive practices as expressions of psychological resistance occasionally appeared as well. In his inspection report on the Dande region of 1928, Jacinto de Sousa thus related a rumour according to which African parents sometimes destroyed male foetuses 'to prevent that there would be *serviçais* [indentured labourers] for the Europeans' (Sousa 1928: 14).[22] But Sousa himself did not know whether to believe this rumour. Most doctors in Angola appear to have assumed, like Damas Mora, that neo-Malthusian practices of birth control were rare, because of the high value the African populations attached to motherhood and parenthood in general. 'When we ponder hygiene and demography', Augusto Ornelas stated, 'we recognize immediately [...] that the births of black children are always wanted, that Malthusian practices are not adopted, that criminal abortions do not exist and, finally, that births are very numerous' (Ornelas 1929: 525; similarly, Amaral 1939: 172). As proof for the innate pronatalism of Africans, doctors and other colonial officials referred to the social advantages of having many children and the consequences of childlessness in African societies. Childlessness was not only believed to be a curse and a dishonour for African women, but it was also a commonly accepted ground upon which their husbands could obtain a divorce, as well as compensation from the woman's family (Silva 1904: 179–180, 185, 254; Cerqueira 1930: 17; Mora 1940: 587–588).

Doctors did not deny that miscarriages and childless women existed, but they generally perceived them to be the consequence of illnesses, most notably of sleeping sickness, yaws and venereal diseases like syphilis and gonorrhoea. And given the rapid recovery of the birth rates in the AMI sectors, leading doctors did not believe that these illnesses would pose a long-term threat to the population's natural ability to reproduce. In their opinion, infertility was, in many cases, a temporary condition that could and would be resolved by treating the disease that had caused it (Silva 1929: 1; Sousa n.d.: 16; Mora 1930: 1; 1940: 594–595). Moreover, many doctors believed that syphilis was 'less damaging to black than to white people', also causing less miscarriages in the tropics than in temperate climates (Mora 1940: 607).[23] For sure, this positive view of African women's fertility was not shared by all, and some doctors wanted to fight the widespread social practices of birth spacing and polygamy to further increase fertility (Sousa 1929: 162).[24] Yet, for most doctors, Angola was clearly not part of what would later be termed the Central African Infertility Belt (Hunt 2007).

Medical demography not only dismissed the notion of low fertility; it simultaneously consolidated the idea that the Angolan population suffered from appalling infant mortality rates and that this was *a*, if not *the* main obstacle to sustained population growth. 'The percentage of children who die in infancy', one AMI doctor summarized it, was so enormous that 'although African women were usually extremely prolific, the population [was] either stable or decreasing instead of increasing' (Gomes 1929: 450). Damas Mora was one of the most tenacious and influential supporters of this view. 'There is, these days, no epidemic scourge that produces victims in a similar or even approximate proportion', he already stressed in 1930 (Mora 1930: 1). Over the next decade, he would repeat this position many times and urge that the colonial health services reconcentrate their efforts on the problem of infant mortality (Mora 1934a: 233; 1940: 559). Clearly, infant mortality had already been accused of being terribly high previously, but it was only in the late 1920s and 1930s that demographic statistics in medical reports turned this assumption into a seemingly objective and irrefutable truth, as they often mentioned infant mortality rates of 50 per cent and more (Freitas and Fonseca 1930: 9; Amaral 1939: 173–174; Teixeira 1931: 21).

Such appalling percentages, which circulated for other colonies as well, were usually the result of a particularly broad definition of infant mortality. Whereas in Europe, 'infant mortality' related to the number of deaths occurring in infants' first year of life, infant mortality rates in colonial Africa seem to have often included all deceased children up to the age of fifteen, and sometimes even miscarriages and stillbirths (see Prum 1927: 18; Thiroux 1931: 569; Van Nitsen 1941: 50–53). In part, this was a consequence of the method used and of ignorance with regard to statistical mathematics and definitions: frequently, as the Belgian doctor Rodhain already criticized at the time (Rodhain 1931: 6, 9, 16), doctors calculated infant mortality rates by simply inverting the total survival index, that is the number of children currently still alive, determined in fertility enquiries (100 per cent minus survival index), without discerning at what age the children had died (Rezende 1929: 311; Mora 1930: 1).

Actually, many a medical report also calculated infant mortality rates according to the international definition (Ornelas and Mesquita 1935: 22; Teixeira 1931: 21; Fonseca 1931: 6–9; Amaral 1935: 37–38, 49–50; 1939: 174). Yet the fact that these more cautious figures, which were usually between 20 and 35 per cent, were less prominently cited, suggests that there was still something else at stake. Indeed, some leading colonial officials preferred to cite the much higher percentages because they served their agenda. Favouring stronger medical interventionism in infant welfare, they were in need of alarming figures to underscore the urgency of the situation and to bolster their demands for investments (Figueira 1938:11; Morna 1944: 174–181). High percentages also served those who wanted to criticize Portuguese colonialism as backward and inhuman (Galvão 1947: 24). And finally, using comprehensive infant mortality rates (0-15 years) also reflected a utilitarian reasoning: the resulting percentage quantified those who did not make it into productive and reproductive life. Significantly, they were sometimes termed '*déchét*' [waste] – those who were lost for the colonial project (Mouchet 1926: 166, 170–171; Prum 1927: 22).

Medical demography, hence, reinforced the idea that infant mortality was much higher in tropical Africa than in Europe. In the Belgian Congo, leading doctors like Mouchet and Rodhain, but also the Governor-General Ryckmans, blamed this on the particularly high mortality during second infancy and adolescence, which they viewed as a distinctive pattern of Central African demography (Mouchet 1926: 170, 173; Rodhain 1931: 16–17; Ryckmans 1933: 257). In a similar vein, Damas Mora believed that mortality rates of children between 1 and 15 years equalled more or less those of infants in their first year of life, thus leading to total rates of 50 to 60 per cent (Mora 1940: 615–616).

Like their foreign colleagues, doctors in interwar Angola did not attribute this divergence to biological differences, but to the civilizational gap that separated both continents. Almost invariably, they blamed high infant mortality primarily on the formidable backwardness and incompetence of African mothers. African mothers, they stated, were still full of prejudices and ignorant about the most basic principles of infant hygiene. They would, for instance, start giving solid food to their infants mere weeks or even days after they were born, thus causing grave digestive disorders, and leave their children naked, thus exposing them to all kinds of diseases. Therefore, it was imperative to replace 'backward' and 'harmful' customs with Western concepts of hygiene and childrearing (Silva n.d.: 9–10; Ornelas 1929).[25]

This racialist argument basically mirrored the accusations that had been previously levelled against lower-class women in Europe (Garrett 1928; Lindner 2014). The solutions would be similar as well. Indeed, like previously in Europe, the first prudent measures of maternal and infant welfare in interwar Angola focused on reforming motherhood through education. At the concentrations and in the few 'native maternities' that were established, African women were taught how to feed their babies correctly and how to protect them against diseases and the dangers of everyday life, a role for which doctors also wanted to train 'itinerant native midwifes' (Ornelas 1929; Silva n.d.: 5–6; Ornelas and Mesquita 1935: 62). In urban settings, so-called 'milk drops' (*lactários*) and dispensaries offered additionally some material assistance, in the form of milk powder, baby clothes and soap (Anonymous 1931).[26]

The cultural explanation for pathologically high infant mortality rates was widespread in the interwar period. This bolsters Megan Vaughan's argument, based on evidence for British East and Central Africa, that culture substituted biology as the central marker of difference in biomedical discourse in interwar Africa (Vaughan 1991: 12–13, 202–203). Yet towards the end of the interwar period, such cultural explanations began to be challenged by a broader debate on the social determinants of health (Packard 2009: 44–48). Thus some doctors and other colonial officials in Angola began to admit that high infant mortality was not only due to culturally determined maternal ignorance, but also to social causes, such as poverty and undernutrition, sometimes exacerbated by colonialism itself, and hence that it was imperative to raise the general standard of life (Amaral 1939: 178, 183; Mora 1940: 605–608).

In conclusion, doctors built upon medical demography to claim that, with few exceptions, the 'native' population in Angola was subjected to a demographic regime that was distinct from the one in Europe in that both fertility and infant mortality were much higher. If, except for the Bushmen case, these differences were interpreted as culturally and socio-economically, and not biologically, determined, that was partly because doctors knew that Europe itself had experienced such a demographic regime in the past and only moved away from it during its industrialization over the last century. Difference, hence, was a matter of change over time. Although the classic version of the demographic transition theory would only be formulated in 1945 by the Princeton-based demographers Notestein and Kingsley, its basic idea that, with the advancement of civilization, all societies would move from a regime of high fertility and (infant) mortality to one of low fertility and (infant) mortality, just like Europe had, was already palpable in interwar population research (Hodgson 1983; Szreter 1993; Mora 1940: 589–594). This is particularly true for infant mortality. By the interwar period, the level of infant mortality was already widely considered to indicate the 'material and moral condition of a nation', that is its level of civilization.[27] Moreover, Portuguese doctors were in a particularly bad position to talk about these differences in terms of biological race if they did not want to discredit their own position. Despite anxieties about falling birth rates, Portugal's birth and infant mortality rates were both still clearly higher than in most of the other European countries in the interwar period (Livi Bacci 1971) and doctors were very aware of this (Garrett 1928; Sousa 1939: 224).

Conclusion

When, in 1940, the first scientific census was conducted in Angola, it borrowed from medical demography by including a simple but large-scale fertility study: all 'native' (that is 'non-civilized' African) women above 14 years were asked by local census agents how many children they had borne and how many of them were still alive (Repartição Técnica de Estatística Geral do Governo Geral de Angola 1940: 64, 68). The results listed 3,035,587 children born from 875,115 mothers, or an average ratio of 3.46 births per woman. For women in the highest age group, this ratio even reached 4.7, with a peak of 5.5 in the Luanda province (where the AMI sectors were located) that seemed to confirm the optimism of doctors like Damas

Mora and Antunes de Amaral (Repartição de Estatística Geral da Colónia de Angola 1941–1947, vol. xi–xii: 107–181, 218).

Interestingly, however, these rates were not calculated on the basis of all adult women, but of mothers only. The more than 360,000 women who had not or not yet borne any children were not taken into account and only briefly mentioned further on, where they were listed according to region and age (ibid.: 217). Certainly, one can only guess about the reasons for this peculiar statistical procedure. But in one of the rare published analyses of the 1940 census data, Alexandre Sarmento, a medical doctor with a strong interest in physical anthropology and demography, conveniently used these inflated rates to bolster his argument about the absence of a natality problem in Angola. Sarmento's argument was also based upon other indices deduced from the census data: a high average birth rate of 35.8 per thousand, the 'progressive' character of the population due to the existence of 40.3 per cent of children, and a low sex ratio of 89.6. Overall, the demographic perspectives of Angola's 'native' population were looking good, he concluded (Sarmento 1948: 644–648).

However, Sarmento also knew that the census was far from perfect and that it did not contain any trustworthy information on such important demographic phenomena as general mortality, infant mortality and stillbirths (ibid.: 635, 649). Hence, although 'scientific' censuses would now be conducted every ten years, they did not entail the end of medical demography. In northern Angola, AMI doctors continued until at least the mid-1950s to register the population and its movements independently from the administrative authorities, thereby attaining natality and mortality rates almost twice as high as the latter (Morgado 1959: esp. 16–24). And in the 1950s and early 1960s, medical doctors including Sarmento still conducted small-scale fertility and infant mortality studies in areas or among ethnic groups that elicited particular interest (Sarmento and Henriques 1954; Ministério do Ultramar – Junta de Investigações do Ultramar 1957: 17–45; Santos 1957: 426–428). Certainly, they lost their virtual monopoly on these kinds of studies: encouraged by students of colonial demography at the Centre of Demographic Studies that had been created in Lisbon in 1944, some administrators in the Portuguese colonies began to conduct small-scale fertility and infant mortality studies as well (see Brito 1953). Yet, overall, the persistence of such sample studies is illuminating. While it testifies to the continuous shortcomings of colonial censuses and civil registries and, through their reliance on doctors and administrators, to the slow professionalization of colonial demography, it also reveals that demographic indices, especially with regard to fertility and infant mortality, continued to be considered markers of both health and difference.

Samuël Coghe is postdoctoral research fellow at the History Department of the University of Giessen. He received his PhD in History from the European University Institute in Florence in 2014 and has been pre- and postdoctoral fellow at the Max Planck Institute for the History of Science in Berlin and at the History Department of the Free University in Berlin. He has published several articles on the history of colonialism in Portuguese Africa and on the abolition of the slave trade and slavery in the nineteenth-century South Atlantic world. He is currently finishing a book manuscript on population politics in Portuguese Angola, while also engaging with a new research project on cattle economies in colonial Africa.

Notes

1. I use these terms in analogy to the distinction between armchair and field anthropologists commonly made in the historiography of anthropology. Whereas armchair anthropologists used to analyse and theorize the information gathered by others (notably missionaries, colonial officials and explorers), field anthropologists would travel and collect their data themselves. Fieldwork became a core condition of social anthropology in the 1920s. On this shift, see Goody (1995).
2. I use 'native' (*indígena*) as an actor's term. While often indiscriminately used during colonial times, *indígena* was basically, much like in the French colonies, a legal term involving a special political and civil statute, used to distinguish between 'uncivilized' and 'civilized' (assimilated) Africans. See, for instance, Silva (2012).
3. See the complaint in Correia (1934: 4–5). An exception is Silva Telles' demographic study on deported Europeans in Angola, see Telles (1903: 38–56). When, in the early 1940s, medical doctor Alexandre Sarmento began to publish on the evolution of the white population in Angola, his analyses were usually based on the data from the general 1940 census. See, for instance, Sarmento (1943).
4. Good overviews of the field of African historical demography can be found in Cordell and Gregory (1987) and Fetter (1990). On newer developments and the challenge posed by cultural and postcolonial studies, see Cordell (2010).
5. Regarding Angola, see, for instance, Heywood and Thornton (1987, 1988).
6. On the history of early French, German and British approaches to sleeping sickness in Africa, see Sarah Ehlers' chapter in this volume and, for the somewhat different Portuguese case, see Coghe (2014: 41–130).
7. For the broader context of bureaucratic rationalization that underpinned these censuses, see Santos (2010).
8. This periodicity followed the recommendations of the International Institute of Statistics and was already implemented in Portugal, like in most European countries, see Sousa (1995: 164–171).
9. Even then, the civil registry did not work well, see Morgado (1954–1955: 113–123).
10. For a similar and very elaborate argument with regard to the Belgian Congo, see Schwetz (1923: 301–312).
11. For a somewhat different method based on 'family tickets', see Amaral (1935: 13–50).
12. On FORÉAMI and its connections with the Angolan AMI programme, see Coghe (2015: 152–154).
13. See, in one of his resumés of this report, Mora (1934b: 39–43). For a more elaborate analysis of this report, see Coghe (2014: 180–181).
14. Some basic notions of demography may have been taught in the courses of public hygiene and epidemiology at the medical schools and the School of Tropical Medicine in Lisbon, however. Personal communication by Rita Garnel, 2 September 2013. See also Garnel (2013).

15. Similar complaints abound for other colonies as well, see, for instance, Bauvallet (1931: 604–605) and Lefrou (1943: 300).
16. For a successful account of inspiring such trust, see Silva (1928: 221–222).
17. On the agency of African nurses in colonial healthcare programmes in general, see also Lyons (1994: 210–213; 221) and Kalusa (2007).
18. For a particularly upfront confession, see Fonseca (1931: 2, 4).
19. In interwar Africa, comparing the distribution of age groups (0-15, adults, old people) in a given population with the standard million (that is the distribution found in the 'standard' population of either Sweden [census of 1890] or England and Wales [census of 1901]) had become a widely used method by colonial demographers to predict whether a given population was regressive (decreasing), stagnating or progressive (increasing). See, for instance, Cazanove (1930: 26–30), Marqueissac (1932: 988), Trolli (1934: 276–279). Kuczynski, for his part, objected that no conclusions with regard to population growth or decline could be drawn from such broad age group ratios (1937: xii).
20. For the debate on the correlation between high sex ratios and population decline in the Pacific, see Widmer (2014).
21. Other doctors also established a causal link between the AMI programme and population growth, see Gama (1929: 540–541) and Amaral (1935: 19–20).
22. Interestingly, this passage was censured in the version that was printed in Lisbon in the 1930s, compare Sousa (n.d.: 17).
23. This belief was probably based on the frequent confusion between syphilis and endemic yaws, see Summers (1991: 787–788) and Headrick (1994: 37–38). See also Silva (1929: 2–3).
24. For such discussions in the neighbouring Belgian Congo, see Hunt (1988).
25. See also Barbara Cooper's chapter in this volume.
26. On the transfer of these *lactários* or *gouttes de lait* from Europe to Africa, see Hunt (1988: 402–406).
27. Quotation from Julia Lathrop, first director of the United States Children's Bureau, in 1921, in Rooke and Schnell (1995: 179). Similarly Sousa (1939: 225).

Archives

AHP: Arquivo Histórico Parlamentar, Lisbon
AHU: Arquivo Histórico Ultramarino, Lisbon
LONA: League of Nations Archives, Geneva

Bibliography

Amaral, A.A. de. 1935. *Relatório da missão médica de reconhecimento nosográfico de Catete, 1930-1931*. Lisbon: Agência Geral das Colónias.

——— 1939. 'Natalidade e mortalidade indígenas', *Boletim Sanitário de Angola* 2: 171–185.

Anonoymous. 1929a. 'Assistência Médica aos Indígenas – Relatório do Chefe da Zona do Congo', *Boletim da Assistência Médica aos Indígenas e da Luta contra a Moléstia do Sono* 3(3): 295–301.

Anonymous. 1929b. 'A obra da assistência ao indígena no Congo', *A Província de Angola*, 16 March.

Anonymous. 1931. 'Protecção à Infância. A , 'Gôta de Leite' é inaugurado em 15 de Agosto', *A Província de Angola*, 8 August.

Appadurai, A. 1993. 'Number in the Colonial Imagination', in C. Breckenridge and P. van der Veer (eds), *Orientalism and the Postcolonial Predicament. Perspectives on South Asia.* Philadelphia: University of Pennsylvania Press, pp. 314–339.

Athayde, L. de M. e. 1918. 'O perigo do despovoamento de Angola', *Boletim da Sociedade de Geografia de Lisboa* 36: 227–243.

Baptista, M.I.R. 2007. 'A demografia em Portugal. Um percurso bibliográfico', *Análise Social* 42(183): 539–579.

Bauvallet, M. 1931. 'Résultats d'une enquête démographique dans le Bas-Dahomey', *Bulletin de la Société de Pathologie Exotique* 24: 604–608.

Bertillon, A. 1882. *Les races sauvages*. Paris: Masson.

Borowy, I. 2009. *Coming to Terms with World Health. The League of Nations Health Organisation, 1921-1946.* Frankfurt am Main: Lang.

Brito, E. 1953. 'Aspectos demográficos dos Balantas e Brames do Território de Bula', *Boletim Cultural da Guiné Portuguesa* 31: 417–470.

Camoesas, J. 1929. 'Sobre a organização da Assistência Médica Indígena', *Boletim da Assistência Médica aos Indígenas e da Luta contra a Moléstia do Sono* 3(2): 140–155.

Carvalho, A.T. de, Governador Interino. 1925. 'Despacho', 16 January, in *Boletim Oficial da Colónia de Angola, Série I*, 17 January.

Cazanove, F. 1930. 'Essai de démographie des colonies françaises', *Bulletin de l'Office International d'Hygiène Publique* 22(8) (supplément).

Cerqueira, I.B. de. 1930. 'Organisação social indigena. Seu estado actual – usos e costumes', in AHU/MU/AGC 2336.

Chesneau, P. 1937. 'Natalité et mortalité infantile en Nord-Annam', *Annales de Médecine et de Pharmacie Coloniales* 35: 688–743.

Clapier, P. 1920. 'Enquête démographique et état actuel de la trypanosomiase au pays Bangala (Afrique Equatoriale Française)', *Bulletin de la Société de Pathologie Exotique* 13: 830–847.

Coghe, S. 2014. *Population Politics in the Tropics. Demography, Health and Colonial Rule in Portuguese Angola, 1890s-1940s*, unpublished PhD thesis. Florence: European University Institute.

——— 2015. 'Inter-imperial Learning and African Health Care in Portuguese Angola in the Interwar Period', *Social History of Medicine* 28(1): 134–154.

Coghe, S. and A. Widmer. 2016. 'Colonial Demography. Discourses, Rationalities, Methods', in Population Knowledge Network (eds), *Twentieth Century Population Thinking. A Critical Reader of Primary Sources.* London/New York: Routledge, pp. 37–64.

Cohn, B.S. 1987. 'The Census, Social Structure and Objectification in South Asia', in B.S. Cohn (ed.), *An Anthropologist among the Historians and Other Essays.* Delhi: Oxford University Press, pp. 224–254.

Cordell, D.D. 2010. 'African Historical Demography in the Postmodern and Postcolonial Eras', in K. Ittmann, D.D. Cordell and G.H. Maddox (eds), *The Demographics of Empire. The Colonial Order and the Creation of Knowledge.* Ohio: Ohio University Press, pp. 22–58.

Cordell, D.D. and J. W. Gregory (eds). 1987. *African Population and Capitalism. Historical Perspectives.* Boulder: Westview Press.

Correia, A.C.G. da S. 1934. 'Os Luso-Descendentes de Angola. Contribuição para o seu estudo antropológico', *III Congresso Colonial Nacional de 8 a 15 de Maio de 1930. Actas das Sessões e Teses.* Lisbon: Tip. e Pap. Carmona.

Costa, A.G. da. 1928a. 'Relatório anual do chefe da zona sanitária do Cuanza Norte (1927)', *Revista Médica de Angola* 6: 21–98.

—— 1928b. 'Relatório sobre a visita aos sectores e postos sanitários', *Revista Médica de Angola* 6: 101–145.

—— 1935. 'Assistência Médica ao Indígena e Combate à Doença do Sono', in F. Mouta (ed.), *Generalidades sobre Angola. Para o 1.° Cruzeiro de Férias às Colónias Portuguesas*. Luanda: Imprensa Nacional, pp. 59–64.

Costa-Sacadura, S.C. da. 1923. *Despopulação em Portugal*. Lisbon: Imprensa Africana.

Curto, J.C. and R.R. Gervais. 2001. 'The Population History of Luanda during the Late Atlantic Slave Trade, 1781-1844', *African Economic History* 29: 1–59.

Dias, A.A. 1933. 'Assistência', *A Província de Angola*, 20 May.

Diniz, J. de O.F. 1914. *Negócios indígenas. Relatório do ano 1913*. Luanda: Imprensa Nacional de Angola.

—— 1918. *Populações indígenas de Angola*. Coimbra: Imprensa da Universidade.

—— 1930. 'Contribuição para o estudo da demografia indígena de Angola', *Boletim da Agência Geral das Colónias* 6(58): 32–53.

Direcção dos Serviços de Saúde e Higiene da Colónia de Angola (ed.). 1941. *Boletim Sanitário de Angola*, referente ao ano 1939. Luanda: Imprensa Nacional.

Ferreira, A.V. 1927. *A situação de Angola*. Luanda: Imprensa Nacional.

Fetter, B. 1990 (ed.). *Demography from Scanty Evidence. Central Africa in the Colonial Era*. Boulder: Lynne Rienner Publishers.

Figueira, M.P. 1938. 'Relatorio do Curador Geral dos Indígenas da Colonia de Angola para 1937', 15 May, in AHU/MU/ISAU 2243.

Fonseca, L.P. de. 1931. 'Relatório do Médico da Sub-Brigada Norte', 30 June, in AHU/MU 585.

Freitas, J.A. de. 1931. 'Relatório do Médico da Sub-Brigada Sul', 31 October, in AHU/MU/DGOPC 737.

—— 1932. 'Relatório do Médico da Sub-Brigada Sul', 29 February, in AHU/MU/DGOPC 737.

Freitas, J.A. de and L.P. de Fonseca. 1930. 'Relatório dos médicos da Brigada de Estradas referente ao reconhecimento de Galangue', 31 December, in AHU/MU/DGAPC 407.

Galvão, Henrique. 1947. 'Exposição do Deputado Henrique Galvão à Comissão de Colónias da Assembleia Nacional', 22 January, in AHP, Section XXVIII, Box 48A, n. 10.

Gama, E.A.D. da. 1928. 'Missão Volante de Assistência aos Indígenas do Planalto de Benguela', *Boletim da Assistência Médica aos Indígenas e da Luta contra a Moléstia do Sono* 2(1-8): 59–66.

—— 1929. 'Um ano de chefia no sector Ambaca-Encoje', *Boletim da Assistência Médica aos Indígenas e da Luta contra a Moléstia do Sono* 3(7-12): 536–541.

Garnel, M.R.L. 2013. 'Da Régia Escola de Cirurgia à Faculdade de Medicina de Lisboa. O Ensino Médico, 1825-1950', in S.C. Matos and J.R. do Ó (eds), *A Universidade de Lisboa, séculos XIX-XX*, vol. 2. Lisbon: Tinta-da-China, pp. 538–650.

Garrett, A. de A. 1928. *Como organizar a luta contra a mortalidade infantil*. Lisbon: Imprensa Nacional.

Gervais, R.R. and I. Mandé. 2007. 'Comment compter les sujets de l'Empire? Les étapes d'une démographie impériale en AOF avant 1946', *Vingtième Siècle. Revue d'Histoire* 95: 63–74.

Gomes, M.G. 1929. 'Reconhecimentos sanitários nas margens dos grandes rios', *Boletim da Assistência Médica aos Indígenas e da Luta contra a Moléstia do Sono* 3(4-6): 443–456.

Goody, J. 1995. *The Expansive Moment. Anthropology in Britain and Africa, 1918-1970*. Cambridge: Cambridge University Press.

Governo Geral da Província de Angola, Secretaria Geral. 1903. *Annuário Estatístico da Província de Angola, 1900 (4°anno)*. Luanda: Imprensa Nacional.

Greenalgh, S. 1996. 'The Social Construction of Population Science. An Intellectual, Institutional, and Political History of Twentieth-Century Demography', *Comparative Studies in Society and History* 38(1): 26–66.

Headrick, R. 1994. *Colonialism, Health and Illness in French Equatorial Africa, 1885-1935*. Atlanta: African Studies Association Press.

Heywood, L. and J. Thornton. 1987. 'Demography, Production, and Labor. Central Angola, 1890-1950', in D.D. Cordell and J.W. Gregory (eds), *African Population and Capitalism. Historical Perspectives*. Boulder: Westview Press, pp. 241–254.

—— 1988. 'African Fiscal Systems as Sources for Demographic History. The Case of Central Angola, 1799-1920', *Journal of African History* 29(2): 213–228.

Hodgson, D. 1983. 'Demography as Social Science and Policy Science', *Population and Development Review* 9(1): 1–34.

Hunt, N.R. 1988. '"Le Bébé en Brousse." European Women, African Birth Spacing and Colonial Intervention in Breast Feeding in the Belgian Congo', *International Journal of African Historical Studies* 21(3): 401–432.

—— 2007. 'Colonial Medical Anthropology and the Making of the Central African Infertility Belt', in H. Tilley and R.J. Gordon (eds), *Ordering Africa. Anthropology, European Imperialism and the Politics of Knowledge*. Manchester: Manchester University Press, pp. 252–281.

Ittmann, K. 1999. 'The Colonial Office and the Population Question in the British Empire, 1918-1962', *The Journal of Imperial and Commonwealth History* 27(3): 55–81.

Jojot, C. 1921. 'Le secteur de la prophylaxie de la maladie du sommeil du Haut-Nyong (Cameroun)', *Annales de Médecine et de Pharmacie Coloniales* 19: 423–442.

Kalusa, W.T. 2007. 'Language, Medical Auxiliaries, and the Re-Interpretation of Missionary Medicine in Colonial Mwinilunga, Zambia, 1922-51', *Journal of Eastern African Studies* 1(1): 57–78.

Kuczynski, R.R. 1937. *Colonial Population*. Oxford: Oxford University Press.

—— 1939. *The Cameroons and Togoland. A Demographic Study*. London: Oxford University Press.

Külz, L. 1910. 'Zur Pathologie des Hinterlandes von Südkamerun, *Archiv für Schiffs- und Tropenhygiene* 14(Beiheft): 1–35.

Lefrou, G. 1931. 'Un nouvel indice de robusticité chez les noirs', *Bulletin de la Société de Pathologie Exotique* 24: 60–67.

—— 1943. *Le noir d'Afrique. Anthropo-biologie et raciologie*. Paris: Payot.

Legrand, E. 1921. 'La dépopulation du Congo belge et les recensements de 1917', *Congo. Revue générale de la colonie belge* 2(1): 201–210.

Lindner, U. 2014. 'The Transfer of European Social Policy Concepts to Tropical Africa, 1900–50. The Example of Maternal and Child Welfare', *Journal of Global History* 9(2): 208–231.

Livi Bacci, M. 1971. *A Century of Portuguese Fertility*. Princeton: Princeton University Press.

Lyons, M. 1994. 'The Power to Heal. African Auxiliaries in Colonial Belgian Congo and Uganda', in D. Engels and S. Marks (eds), *Contesting Colonial Hegemony. State and Society in Africa and India*. London: I.B. Tauris & Co, pp. 202–223.

Marqueissac, de. 1932. 'Enquête et sondages démographiques en pays Kabré (Nord-Togo)', *Bulletin de la Société de Pathologie Exotique* 25: 986-992.

Matos, J.N. de. 1923. 'Portaria Provincial n. 148', 6 August, in *Boletim Oficial da Província de Angola, Série I*, 11 August: 314–316.

—— 1926. *A Província de Angola*. Porto: Edição de Maranus.

Matos, P.F. de. 2006. *As Côres do Império. Representações Raciais no Império Colonial Português*. Lisbon: Imprensa de Ciências Sociais.

Mercier, Dr. 1926. 'Conférence sur la démographie et les statistiques démographiques au Togo, faite en présence de la mission des médecins échangistes à Atakpame', 20 April, in LONA, R 955, 12B/54511/41908.

Ministério da Marinha e Ultramar. 1900. 'Lei auctorisando o governo a proceder de dez em dez anos ao recenseamento geral da população nas populações ultramarinas portuguezas', in *Collecção Official da Legislação Portuguesa anno de 1899*. Lisbon: Imprensa Nacional, p. 343.

Ministério do Ultramar – Junta de Investigações do Ultramar (ed.). 1957. *Contribuição para o estudo da fertilidade da mulher indígena no Ultramar português*. Lisbon: Junta de Investigações do Ultramar.

Mora, A.D. 1930. 'Um belo gesto das senhoras de Luanda', *A Província de Angola*, 1 August.

——— 1934a. *A luta contra a moléstia do sono em Angola (1921-1934)* (Relatórios da Direcção dos Serviços de Saúde e Higiene de Angola 2), Luanda.

——— 1934b. 'O estado actual da Assistência Médica aos Indígenas na colónia de Angola e outras colónias estrangeiras do grupo da Africa inter-tropical', in *III Congresso Colonial Nacional de 8 a 15 de Maio de 1930. Actas das Sessões e Teses*. Lisbon: Tip. e Pap. Carmona.

——— 1934c. 'Assistência Médica aos Indígenas, em Angola', *A Província de Angola*, 6–8 August.

——— 1940. 'A mortalidade infantil de brancos e indígenas nas Colónias de Angola e Moçambique, suas causas principais e remédios possíveis. Métodos para a organização de estatística da mortalidade infantil', in Comissão Executiva dos Centenários (ed.), *Memórias e comunicações apresentadas ao Congresso Colonial (IX Congresso)* (*Publicações do Congresso do Mundo Português* 14). Lisbon, pp. 557–625.

Morgado, N.A. 1954–1955. 'A demografia do Ultramar português. Estudo descritivo e crítico da posição actual no que se refere a documentação estatística e estudos relativos à demografia ultramarina', *Revista do Centro de Estudos Demográficos* 9: 71–283.

——— 1959. *Aspectos da evolução demográfica da população da antiga província do Congo (1949-1956)*. Lisbon: Centro de Estudos Políticos e Sociais.

Morna, A. de F. 1944. *Angola. Um ano no Governo Geral (1942-1943), 1. Volume*. Lisbon: Livraria Popular de Francisco Franco.

Mouchet, R. 1926. 'La natalité et la mortalité infantile dans la Province Orientale', *Annales de la Société Belge de Médecine Tropicale* 6: 165–174.

Ornelas, A. 1929. 'A obra de protecção à criança negra', *Boletim da Assistência Médica aos Indígenas e da Luta contra a Moléstia do Sono* 3(7-12): 523–526.

Ornelas, A. and B.P. de Mesquita. 1935. *Relatório da missão médica de assistência aos indígenas do Cuanza, 1929*. Lisbon: Agência Geral das Colónias.

Overath, P. 2011. 'Bevölkerungforschung transnational. Eine Skizze zu Interaktionen zwischen Wissenschaft und Politik am Beispiel der "International Union for the Scientific Study of Population"', in P. Overath (ed.), *Die vergangene Zukunft Europas. Bevölkerungsforschung und -prognosen im 20. und 21. Jahrhundert*. Cologne, Weimar and Vienna: Böhlau, pp. 57–83.

Packard, R. 2009. 'The History of the Social Determinants of Health in Africa', in H.J. Cook, S. Bhattacharya and A. Hardy (eds), *History of the Social Determinants of Health. Global Histories, Contemporary Debates*. Hyderabad (India): Orient Black Swan, pp. 42–77.

Peabody, N. 2001. 'Cents, Sense, Census. Human Inventories in Late Precolonial and Early Colonial India', *Comparative Studies in Society and History* 43: 819–850.

Peiper, O. 1920. *Geburtenhäufigkeit, Säuglings- und Kindersterblichkeit und Säuglingsernährung im früheren Deutsch-Ostafrika*. Berlin: Schoetz.

Piot, A. 1920. 'Sur le fonctionnement d'un secteur de prophylaxie contre la trypanosomiase au Congo français (1919)', *Bulletin de la Société de Pathologie Exotique* 13: 376–384.

Prum, T. 1927. 'Observations concernant la natalité et la mortalité infantile de la région de Leverville', *Annales de la Société Belge de Médecine Tropicale* 7: 15–22.

Reis, C. dos S. and A. Sarmento. 1960. *Manual de Estatistica médica.* Lisbon: Instituto de Medicina Tropical.

Repartição Central de Estatística Geral de Angola. 1936. *Anuário estatístico de Angola, Ano de 1934.* Luanda: Imprensa Nacional.

Repartição de Estatística Geral da Colónia de Angola (ed.). 1941–1947. *Censo Geral da População, 1940,* 12 vols, Luanda: Imprensa Nacional.

Repartição Técnica de Estatística Geral do Governo Geral de Angola. 1940. *Bases para a execução do censo da população da Colónia em 1940.* Luanda: Imprensa Nacional.

Rezende, A. de. 1929. 'Missão volante da Lunda', *Boletim da Assistência Médica aos Indígenas e da Luta contra a Moléstia do Sono* 3(3): 310–311.

Ribeiro, A.P. 1933. *Apontamentos para o estudo da vitalidade das populações cafreais de Angoche. Assistência médica ao indígena durante o ano de 1932.* Lourenço Marques: Imprensa Nacional.

Ricon-Ferraz, A. and H. Guimarães (eds). 2008. *Sobre a mortalidade infantil no Porto e as formas de a evitar. Facsimile da Dissertação de concurso do Professor António de Almeida Garrett.* Porto: Universidade do Porto.

Rivers, W.H.R. 1922. 'The Pyschological Factor', in W.H.R. Rivers (ed.), *Essays on the Depopulation of Melanesia.* Cambridge: Cambridge University Press, pp. 84–113.

Rodenwaldt, E. 1915. 'Ein Beitrag zu der Frage des Bevölkerungsrückgangs in den afrikanischen Schutzgebieten', *Mitteilungen aus den deutschen Schutzgebieten* 28(3): 145–160.

Rodhain, J. 1931. *La mortinatalité et la mortalité infantile au point de vue pathologique – Afrique Centrale* (Reports to the International Conference on African Children 4). Geneva: UISE.

Rooke, P.T. and R.L. Schnell. 1995. '"Uncramping Child Life". International Children's Organisations, 1914-1939', in P. Weindling (ed.), *International Health Organisations and Movements, 1918-1939.* Cambridge: Cambridge University Press, pp. 176–202.

Roque, R. 2001. *Antropologia e império. Fonseca Cardoso e a expedição à Índia em 1895.* Lisbon: Imprensa de Ciências Sociais.

——— 2010. *Headhunting and Colonialism. Anthropology and the Circulation of Human Skulls in the Portuguese Empire, 1870-1930.* Basingstoke and New York: Palgrave Macmillan.

Rosental, P.-A. 2007. 'Wissenschaftlicher Internationalismus und Verbreitung der Demographie zwischen den Weltkriegen', in P. Krassnitzer and P. Overath (eds), *Bevölkerungsfragen. Prozesse des Wissenstransfers in Deutschland und Frankreich (1870 - 1939).* Cologne: Böhlau, pp. 255–291.

Ryckmans, P. 1933. 'Démographie Congolaise', *Africa. Journal of the International African Institute* 6(3): 241–258.

Sant'Anna, J.F. 1923. 'O problema da assistência médico-sanitária ao indígena em Africa', *Revista Médica de Angola* 4(2): 73–178.

——— 1931. *1.° Relatório da Missão Médica ao Arquipélago de Cabo Verde em 1930.* Typescript consulted in the Library of the Instituto de Higiene e Medicina Tropical in Lisbon.

Santos, A.C.V.T. dos. 1945. *Angola, coração do império,* Lisbon: Agência Geral das Colónias.

Santos, C.M. 2010. 'Administrative Knowledge in a Colonial Context. Angola in the Eighteenth Century', *The British Journal for the History of Science* 43(4): 539–556.

Santos, D.R. dos. 1957. 'Aproveitamento dos postos experimentais e de controle para educação alimentar e melhoria da alimentação da população controlada', in *Conferência inter-*

africana de nutrição. 3. sessão, Luanda, Outubro 1956. Communicações, vol. 2. London: C.C.T.A., pp. 425–440.

Santos, F.F. dos. 1923. 'Assistência Médica aos Indígenas e processos práticos da sua hospitalisação', *Revista Médica de Angola* 4(2): 51–71.

Santos, G. 2012. 'The Birth of Physical Anthropology in Late Imperial Portugal', *Current Anthropology* 53(S5): 33–55.

Sarmento, A. 1943. 'O estudo da população de Angola', *Occidente. Revista Portuguesa de Cultura* 19(60): 419–424.

——— 1944. *Aspectos da natalidade e mortalidade infantil em Angola. Separata do Jornal do Médico*. Porto: Costa Carregal.

——— 1948. 'População indígena de Angola (Sondagens e perspectivas demográficas)', *Boletim da Sociedade de Geografia de Lisboa* 66: 635–649.

Sarmento, A. and F.F. Henriques. 1954. 'Alguns aspectos demográficos dos Bochimanes do Sul de Angola', *O Médico* 5(149): 567–572.

Schiebinger, L. 2004. 'Feminist History of Colonial Science', *Hypatia* 19(1): 233–254.

Schwetz, J. 1923. 'Contribution à l'étude de la démographie congolaise', *Congo. Revue générale de la colonie belge* 4(1): 297–340.

——— 1924. 'Deuxième contribution à l'étude de la démographie congolaise', *Congo. Revue générale de la colonie belge* 5(1): 333–365.

Silva, A.M. da. n.d. *Serviço de Assistência aos Indígenas no distrito do Congo, 1930*, Lisbon: Agência Geral das Colónias.

Silva, C.N. da. 2012. 'Natives who were 'Citizens' and Natives who were *Indígenas* in the Portuguese Empire (1900-1926)', in A.W. McCoy (ed.), *Endless Empire. Spain's Retreat, Europe's Eclipse, America's Decline*. Madison: University of Wisconsin Press, pp. 295–305.

Silva, F.V. da. 1928. 'Relatório do Chefe da Missão Volante do Distrito de Malange referente aos meses de Setembro a Novembro de 1927', *Revista Médica de Angola* 6: 219–243.

——— 1929. 'O pian', *Boletim da Assistência Médica aos Indígenas e da Luta contra a Moléstia do Sono* 3(1): 1–16.

——— 1936. *Relatório do Serviço Permanente de Prevenção e Combate à Peste Bubónica no Sul de Angola*, 1933. Lisbon: Agência Geral das Colónias.

Silva, J. de M. e. 1904. *Contribuição para o estudo da região de Cabinda. Memória para o Congresso Colonial Nacional*. Lisbon: Typographia Universal.

Soloway, R.A. 1990. *Demography and Degeneration. Eugenics and the Declining Birthrate in Twentieth Century Britain*. Chapel Hill: University of North Carolina Press.

Sousa, C.S. de. 1939. 'Necessidades e deficiências da assistência infantil', *Revista Portuguesa de Pediatria e Puericultura* 2(5): 221–242.

Sousa, F.A.P. 1995. *A História da Estatística em Portugal*. Lisbon: INE.

Sousa, J. de. 1928. 'Relatório da Missão Médica Volante de Assistência aos Indígenas do Dande, 1928', in AHU/MU/AGC 2336.

——— n.d. *Relatório da missão médica volante de assistência aos indígenas do Dande, 1928*. Lisbon: Agência Geral das Colónias.

——— 1929. 'O capital humano', *Boletim da Assistência Médica aos Indígenas e da Luta contra a Moléstia do Sono* 3(2): 160–162.

Summers, C. 1991. 'Intimate Colonialism. The Imperial Production of Reproduction in Uganda, 1907-1925', *Signs* 16(4): 787–807.

Szreter, S. 1993. 'The Idea of Demographic Transition and the Study of Fertility Change. A Critical Intellectual History', *Population and Development Review* 19(4): 659–701.

Teixeira, W.G. 1931. 'Relatório da Zona Sanitária do Cuanza para 1930', Junho de 1931, in AHU/MU/AGC 2336.

Telles, F.X. da S. 1903. *A transportação penal e a colonização*. Lisbon: Typ. Liv. Ferin.

Thiroux, A. 1931. 'La natalité et la mortalité infantiles dans les colonies françaises', *Revue Philantropique* 51(408): 561–569.

Tomlinson, R.P. 1985. 'The "Disappearance" of France, 1896-1940. French Politics and the Birth Rate', *The Historical Journal* 28: 405–415.

Trolli, G. 1934. 'Contribution à l'étude de la démographie des Bakongo', *Bulletin des Séances de l'Institut Royal Colonial Belge* 5(2): 239–316.

Trolli, G. and L. Dupuy. 1934. *Contribution à l'étude de la démographie des Bakongo au Congo Belge 1933*. Brussels: M. Cock.

Valentim, A. 2000. *Velho Brasil, Novas Áfricas. Portugal e o Império (1808-1975)*. Porto: Ed. Afrontamento.

Van Nitsen, R. 1941. *Contribution à l'étude de l'enfance noire au Congo Belge*. Brussels: van Campenhout.

Vaughan, M. 1991. *Curing their Ills. Colonial Power and African Illness*. Cambridge: Polity Press.

Widmer, A. 2012. 'Of Field Encounters and Metropolitan Debates. Research and the Making and Meaning of the Melanesian "Race" during Demographic Decline', *Paideuma* 58: 69–93.

——— 2014. 'The Imbalanced Sex Ratio and the High Bride Price. Watermarks of Race in Demography, Census, and the Colonial Regulation of Reproduction', *Science, Technology & Human Values* 39: 538–560.

9

Indo-Europeans in the Dutch East Indies

An Indo-European Analysis of a Paradoxical Colonial Category

Hans Pols

During the first four decades of the twentieth century, discussions about the 'Indo problem' became increasingly common in the Dutch East Indies.[1] Even though most social commentators appeared to have very clear ideas about who belonged in this group and who did not, it is not easy to define the criteria for inclusion that were implicit in these discussions. In principle, individuals were defined as Indo-Europeans when they had both European and Indonesian ancestry; in practice, however, individuals were labelled as such when they did not easily fit into prevalent racial categories: they were either Europeans with more Indonesian characteristics than most upper-middle-class Europeans were comfortable with or Indonesians with more European characteristics than could be easily ignored. Compounding the difficulties in defining this group and identifying its members, the category 'Indo-European' was not an official category used by the colonial administration. The category 'Indo' (the Dutch phrase generally used to indicate Indo-Europeans) did not therefore exist in the official statistics that were routinely collected by civil servants. The offspring of Europeans and Indonesians were either classified as European or as native, depending on the circumstances of their birth. In everyday life, the category 'Indo-European' was highly flexible and its boundaries malleable and permeable. Elite families with mixed indigenous and European ancestry were generally not considered Indo-European at all; because of their social status and economic success, they were seen as European *pur sang*. In contrast, impoverished Europeans who had moved to the city's indigenous quarters and had taken on an indigenous lifestyle (irrespective of whether they had any Indonesian ancestry) were commonly considered 'Indo-European' or, at times, even 'native'. Social status, economic success and poverty ordinarily served as a proxy for admixture with native blood.

In both official investigations on health, social status, income and poverty among Europeans and broader public discussions about the Indo problem, racial, social and cultural definitions at times appeared to overlap while at other times they contradicted each other. This predicament reflected the indeterminate, flexible and malleable nature of racial typologies commonly used in the Indies as well as the ambiguous, paradoxical and continuously contested spaces Indo-Europeans occupied in colonial society. In Ann Laura Stoler's words, the colonial discourse on this group reveals the 'construction of colonial categories and national identities' and indicates the precarious situation of 'those people who ambiguously straddled, crossed, and threatened these imperial divides' (2002a: 79). The multiple registers in which Indo-Europeans were implicitly and explicitly defined demonstrates that, in the Dutch East Indies, definitions of race were inherently related to broader social, cultural and ethnic considerations held by the general public, colonial administrators and investigators as well as the way in which these groups constructed, or denied, difference in colonial societies. Medical researchers concerned with categories of difference or mixture could view health as related to social and cultural issues like poverty and sexual impropriety, or relate them to inherent racial and ethnic categories, leading to radically different approaches to these conditions. Alexandra Widmer has commented, in her research on racial classification in a different context, on 'the malleability of racial thinking and the intransigence of racial categorization' (2012: 69). Despite the flexibility with which racial categories were used in the Dutch East Indies and their malleability to suit particular purposes, these categories themselves remained intact (although, of course, their exact definitions changed over time). In effect, one could say, 'race' was never just about race.

In this chapter, I explore one of the very few academic investigations by an Indo-European on Indo-Europeans because it illuminates, in a way that was unusual at the time, the arbitrary, haphazard and at times paradoxical nature of discussions around Indo-Europeans in the Dutch East Indies. In 1931, economist Joseph Theodore Koks, who was born near Batavia (now Jakarta) in 1902, received his doctorate at the University of Amsterdam for his sociological investigation of Indo-Europeans in the Dutch East Indies. An extended version of his dissertation (Koks 1931) appeared one year later as a book entitled *De Indo* (The Indo-European).[2] Koks' book makes for fascinating reading. Although Koks provided an elaborate definition of 'Indo-European' with specific criteria for inclusion, his primary aim was to provide a sociological analysis of several different sub-groups of Indo-Europeans. To many readers today, most of his conclusions will not be too surprising; at the time his work was published, however, they were novel. In my opinion, Koks' definitions and conclusions are not the most interesting part of his work; it is far more interesting to read his book as an elaborate deconstruction of general, social, cultural and biological definitions of Indo-Europeans and, thereby, race in a colonial context. Time and again, Koks illustrated the futility of defining Indo-Europeans and continuously destabilized and undermined common ideas, anthropological thinking, medical theories and political rhetoric about this group. As a group of in-between individuals, Indos repeatedly upset colonial boundaries and incessantly defied official categorization and theoretical explanation. Koks fully embraced the disruptive potential inherent in the social position of Indo-Europeans in the Dutch East Indies by writing a delightful academic satire that disrupted, challenged and undermined colonial categories. Ultimately, Koks deconstructed a colonial paradox.

Legal, Anthropological, Social and Cultural Views on Indo-Europeans

The legal code of the Dutch East Indies was based on the racial classification of its inhabitants in three categories: European, foreign Oriental and 'native' (in Dutch: *inlander*) (see Fasseur 1994).[3] Individuals were defined as European when they were born in Europe or were the legitimate children of fathers with European legal status. The offspring of European men and native women were legally classified as European when the parents were married at the time of birth or when the father legally accepted his children by adopting them; otherwise, they were classified as native. Consequently, the legal status of children of mixed parentage depended on a number of accidental and fortuitous factors which were unrelated to biological ones. The category 'foreign Oriental' applied to the Chinese and Arab inhabitants of the Indies who mostly lived in the urban centres and engaged in trade and ran shops. In 1898, after a trade treaty was signed with Japan, the Japanese inhabitants of the Indies were included in the European category as well. The basic definitions of the Dutch East Indies legal code preceded all other definitions and categorizations that were used by the colonial administration. All information collected by bureaucrats on the inhabitants of the Indies, and thereby all statistical information that was generated on the basis of this data, therefore used this legal categorization. In the way in which these categories were interpreted, however, biological, anthropological and broader social definitions were often implicitly included.

The categories in the Dutch East Indies legal code were permeable to a modest degree: individuals classified as 'foreign Oriental' or 'native' could formally apply to be considered 'equivalent to European' if they were able to demonstrate that they lived in a European milieu, spoke Dutch, were educated, maintained a European lifestyle (and could afford to do so), and had relinquished their ties to their ethnic group. According to these rules, being 'European' was defined on the basis of education, income and social milieu rather than ethnicity or race.[4] However, legal classification as European did not easily translate into acceptance in European social circles. Natives who were declared 'equivalent to European' were not welcome in European clubs and were not treated as Europeans in Batavia's better neighbourhoods. Individuals of mixed ancestry who conformed to the social expectations of upper-class European society and were therefore not considered Indo-European, on the contrary, were warmly welcomed. The opposite was not uncommon either: when poor Europeans moved to indigenous neighbourhoods and lived like natives, they were considered 'dissolved in the *kampung* (native quarters)' as the legal jargon of the day had it, and were not welcome in European circles. Categorization on the basis of the Dutch East Indies legal code was not always as rigid as it appeared; at the edges, it had a modicum of flexibility, allowing a select few natives to ascend to European status and condemning impoverished Europeans to a reverse trajectory. The ascent or descent within the colonial social structure of the Dutch East Indies was commonly interpreted in racial or ethnic terms, indicating how broadly they were defined in everyday life.

Before the turn of the twentieth century, the majority of European inhabitants in the Indies were of mixed descent for the relatively straightforward reason that virtually all European migrants to the Indies were male. In 1859, of the 27,330 Europeans living in the Indies, 23,086 were born there (Bosma 1989: 20). It is more difficult to ascertain the total number of people living on Java; some sources estimate Java's population to be around 12 million (The Population of Indonesia 1974: 6). The unions of newly arrived European men with Indonesian women created a group of mixed-race individuals who, over several generations, formed their own community; their members occupied most administrative and trade positions in the colonies. In Central Java, some Indo-European families became extremely wealthy by marrying into the families of the local sultans and by establishing plantations. As Jean Gelman Taylor has pointed out, before the 1870s, newly arrived European men with social ambitions married women from prominent Indo-European families to enhance and solidify their social status (Taylor 1983).[5] The category 'Indo-European' was not used at this time; it was meaningless as it would have contained almost all Europeans living in the colonies (only excluding newly arrived migrants and individuals holding the highest administrative positions).

This situation began to change in the last three decades of the nineteenth century. After the Suez Canal was opened in 1869, travel time from the Netherlands to the Indies was reduced from months to weeks. Around the turn of the twentieth century, discoveries in tropical medicine greatly reduced anxieties about health and disease in the tropics (see Monnais and Pols 2014). At the same time, an ever-growing number of European professionals arrived in the Indies to work in the expanding plantation economy and growing colonial bureaucracy. These new migrants often arrived with their families and intended to remain in the colonies for a limited time only.[6] They occupied the highest and most remunerative positions, sent their children to the best schools, which followed the Dutch curriculum, and deliberately stayed aloof from the established European population in the Indies. This new group of temporary migrants to the Indies were generally called *trekkers* (Dutch: migrants), as opposed to those Europeans who had lived in the Indies for generations, who were now designated *blijvers* (Dutch: 'stayers'). The new European migrants were also derisively called *totok* (Malay: 'full-blood') European; at the same time, the moniker 'Indo' came to designate poor, uneducated and undesirable Europeans who bore an uncomfortably close resemblance to the indigenous population in character, manners, lifestyle and appearance. As a consequence of changed migration patterns, the European community in the Indies became increasingly fractured. Even though members of the European community of old, newly arrived Europeans and Indo-Europeans were all European from a legal perspective, social differentiation within the European social sphere became more pronounced after the turn of the twentieth century. In everyday life, being European was defined by a variety of social and cultural characteristics, including the occupation and income of the head of the household, the education both parents had received, the character, style and cultural refinement of the wife, the school the children were attending, the nature of the family dwelling, lifestyle and consumption patterns, the quality and flawlessness of the Dutch they spoke, the company the family kept, the type of food they ate, as well as physical appearance and style of dress.

The highest positions in the colonial bureaucracy and the most rewarding occupations in the plantation economy were open only to individuals who held European qualifications. Most totok Europeans but only a small number of individuals from long-established families in the Indies were able to send their children to the Netherlands for their education, which also came with the guarantee that their children would acquire a fully European lifestyle and speak accent-free Dutch. For the mixed-race individuals among them, this guaranteed their status as fully European – once they returned to the colonies, they could leave their Indo-European identity behind and pass as *totok* or white. Yet despite the satisfaction of leading a privileged life in the colonies, a family's European status was in principle always uncertain and potentially under threat. A sudden loss of income, misadventure, failed investments, crop failures or unexpected market fluctuations could potentially undermine its social standing and could, eventually, lead to the loss of the coveted *totok* European status. Individuals who prided themselves on their European status firmly distinguished themselves from those who bore, in any possible way, the marks of poverty or financial decline, a prolonged stay in the Indies, or an affinity (or worse, relatedness) to the indigenous population. Individuals whose social rank declined and who socially became increasingly marginalized were seen as Indo-European.

During the last decades of the nineteenth century, Dutch politicians and colonial administrators had expressed deeply ambivalent opinions about the fate of indigent Europeans in the colonies. On the one hand, administrators thought that poor Europeans constituted a blemish on Europe's superior reputation in the Indies; on the other hand, they felt some measured compassion towards Europeans who had fallen on hard times. At the same time, the issue of poverty was increasingly viewed as a specifically Indo-European problem. Poor Europeans tended to move to indigenous quarters and acquire a more or less indigenous way of life. Because 'Indo-European' was not an official category used by the colonial administration, it did not have any statistical information on individuals labelled that way – hence, the several official investigative committees devoted to addressing the problem of poverty among Europeans had to ascertain who, of all the individuals they encountered, rightfully fitted this category and who did not, which caused endless problems as it often required asking probing questions of a deeply personal and potentially embarrassing nature, such as whether one's parents were legally married, or whether one currently lived with a concubine or a spouse. These inquiries were not welcomed by the individuals who became the object of official interest and investigation, as they probed deeply embarrassing and potentially stigmatizing issues.

Stoler has analysed the reports of several commissions set up to investigate the causes of poverty among Europeans (2009: 141–178).[7] Investigators for the 1902 poverty commission, for example, asked a series of probing questions about living conditions, marital status, employment history, moral standards, and interactions with natives. In the end, it was concluded that the poor individuals who had been the topic of investigation were not European at all; labelling them that way was a 'misnomer, a category error, a misapprehension' (ibid.: 173). According to the members of the poverty commission, this miscategorized group consisted, in fact, of mixed-race individuals who had reverted to their indigenous roots. Because they were natives for all intents and purposes, the problem under investigation did not, in fact, exist. One could call the conclusions of the poverty commission a mere semantic play; by re-categorizing the population which had fallen into abject poverty as native (the label 'Indo-European' merely indicated that they had commenced on a social downward trajectory), their living conditions could now be re-described as average or, at least, not unusual.[8] In Dutch East Indies social life, racial categories were defined in multiple registers – by ethnic origin, descent, social status, milieu, everyday habits and moral character. In both everyday social life and in official discussions, these categories were ambiguous, flexible and applied haphazardly. In public discourse, the category 'Indo-European' functioned as an expression of apprehension, disapproval, reproach and, ultimately, moral condemnation. Yet despite the interpretive flexibility of racial/legal categories in everyday life and official discourse, these categories themselves remained rigid and not amenable to change.

As social historian Ulbe Bosma (2005) has argued, the influx of a new type of European migrant after the turn of the twentieth century led to broader social concerns about Indo-Europeans and, later, to the formation of a group which self-consciously adopted this designation. Because the Dutch East Indies had limited educational opportunities, most Indo-Europeans came to occupy the lowest ranks in the colonial administration, businesses and plantations. At the same time, educational opportunities for Indonesians expanded, which enabled them to replace Indo-Europeans in their low-level jobs as they were much cheaper to employ. Europeans could not own land in the colonies as this right was reserved for the indigenous population; Indo-Europeans were thus unable to pursue careers in agriculture. Seeing the number of positions to which they could apply dwindle, Indo-Europeans felt increasingly marginalized in colonial society; they found it difficult to make ends meet, and were relegated to the least desirable European neighbourhoods. This situation led, in 1919, to the founding of the *Indo-Europeesch Verbond* (IEV; Indo-European Association), which advocated for Indo-Europeans in the *Volksraad* (Dutch: People's Council), the newly established colonial parliament.[9] In the 1920s and 1930s, it increasingly identified with conservative and even reactionary political groups. Members emphasized the European legal status of Indo-Europeans and minimized the cultural, social and ethnic differences between them and *totok* Europeans. Yet with respect to the native population, Indo-Europeans strongly emphasized racial differences. Even among Indo-Europeans, definitions of what it meant to be Indo-European were ambiguous and, at times, paradoxical.

In Dutch colonial novels, life in the Dutch East Indies is generally not portrayed favourably. Famous (or notorious) are Bas Veth's elaborate depictions, published in 1900, of the miserable life in the tropics and the general degeneration among Europeans there:

> For me, the Dutch East Indies are the incarnation of misery. The twelve years I spent in this region of exile are twelve dire dreams to me. I have not found anything uplifting there; everything was gloomy. Good people turn bad there; the reverse never occurs. … [Everywhere,] the same phenomena: degeneration and the decomposition of the pure European character. It is an insidious and treacherous process, which remains unnoticed, like first-stage cancer, like beriberi, like diabetes.
>
> The cries of contempt and disgust aroused by so many acts of Europeans in the East turn into sounds of compassion and pity when one understands how sick those Europeans are.
>
> Yes, the archipelago is one [large] hospital for sufferers of degeneration. (Veth 1900: 1–2 [author's translation])[10]

According to Veth and several other authors of Dutch colonial novels, the most important factor leading to the poor health of Europeans, both physical and mental, was the tropical climate. Many diseases common in the Indies – cholera, plague, tuberculosis, anaemia, dysentery, cancer – adorn his text.[11] Not only was the climate conducive to the spread of disease and degeneration in general, it also inspired Europeans to relinquish the sexual proprieties they had maintained at home, leading to the intrigues and depravity that were part and parcel of colonial novels.

In addition to the sick and anaemic state of most Europeans in the Indies, the atrophied nature of social life, the transient nature of European communities, the generous consumption of alcoholic beverages, the frequent use of obscene language, the omnipresence of gossip and rumour, and the shallow desire to make money quickly, Veth also condemned the frequent marriages between European men and mixed-race (or even indigenous) women. He was convinced that such marriages could not succeed because 'races, so different in nature, views, lifestyle, constitution […] cannot fit together with any chance of happiness and harmony' (Veth 1900: 147). According to Veth, Indo-European women were still steeped in the indigenous world of magic, spells, spirits, incantations and herbal potions; as a consequence, Europeans and Indo-Europeans 'are worse than strangers: they are mental enemies' (ibid.: 160).[12] Veth was unusually peevish about European men marrying their native concubines.[13] He claimed that a civil war between Europeans and Indo-Europeans was inevitable and would spell the end of the Dutch East Indies.

In Dutch colonial literature, Indo-Europeans are often portrayed in derogatory terms as indolent, lazy, sensuous, gullible, unreliable, deceptive and, at times, even dangerous (the indigenous populations was often described in the same way) (Alatas 1977). In Dutch naturalist novels about colonial life, tales of degeneration are not uncommon.[14] In lighter forms of literature such as newspaper serials and plays, Indo-Europeans often figured for comic effect.[15] In particular, their broken Dutch spoken with a strong Malay inflection unfailingly served that purpose.

Indo-European novelist and literary critic Alfred Birney (2012) has recently argued that most of the best-known literary works about the Dutch colonial life were written by individuals who were only superficially acquainted with colonial life. In particular, he is critical of their portrayal of Indo-Europeans and their lack of differentiation of subgroups of Indo-Europeans. By exploring literary works written by Indo-Europeans rather than European visitors to the colonies, Birney cogently demonstrates how these novels provide nuanced insights into the multi-layered and highly differentiated cultural and social life of Indo-Europeans. By analysing these novels, Birney explores how Indo-Europeans themselves experienced and interpreted their interactions with recently arrived Europeans, with Indonesians and with other Indo-Europeans. In particular, he focuses on how Indo-Europeans interpreted the use of cultural symbols, attitudes, expressions and ways of behaving among Indo-Europeans by relating these to specific backgrounds, social positions and broader aspirations. Although Birney and Koks wrote at very different times and their social situations were very different, both authors identified as Indo-European and wrote about Indo-Europeans. Like Birney, Koks repeatedly emphasized the need to differentiate between different types of Indo-Europeans, although he did so some eight decades earlier.

Koks and the Deconstruction of Theories and Views on Indo-Europeans

Joseph Theodore Koks explicitly defined himself as Indo-European although most Europeans living in the colonies would classify him as European. Koks was born in 1902 in Meester Cornelis (today's Jatinegara), a village just southeast of Batavia that housed a large army encampment and had a large Indo-European community. Koks came from a respectable family.[16] His grandfather was an economist employed by the colonial Department of Finance and his father was a pharmacist who worked for Batavia's opium factory; he became a lecturer at the Batavia Medical School after attaining a doctorate in medicine in the Netherlands (Koks 1922). Koks grew up with his grandparents, his parents and his uncle's family. His uncle, like his grandfather, worked for the Department of Finance. Also living with the extended Koks family was the widow of Koks' great-grandfather, who had been a former civil servant in the lower echelons of the colonial bureaucracy and had worked on various locations in the Indies. His widow was an Indonesian woman from the highlands near Padang, Sumatra. Koks' great-grandfather was the son of a corporal in the colonial army who had also married an Indonesian woman from the same area. From his childhood, Koks was therefore familiar with Indo-Europeans with a different social status and career trajectories, which is reflected in his writings.

From 1916 to 1922, Koks attended the most prestigious high school in the Indies, Batavia's *Koning Willem III* School. In 1924, he moved to the Netherlands to pursue an advanced education in economics. He wrote his dissertation under the guidance of S.R. Steinmetz, a professor of ethnology and sociology at the University of Amsterdam who was deeply influenced by social Darwinism and was a moderate eugenicist (Pols 2010). During his sojourn in the Netherlands, Koks was active in several student groups and an association for Indo-Europeans, and married an Indo-European woman.[17] After obtaining his doctorate, he returned to the Indies and was offered a position at the Department of Finance where he remained until the Japanese armed forces occupied the Indies in 1942.[18] Given Koks' family background, his studies in the Netherlands, and his career trajectory, it is clear that he came from a well-to-do family who was participating in the migratory circuit of educated, upper-middle class Europeans. Because of his social standing, he and his family would not be considered Indo-European at all. One of the aims of Koks' dissertation was to broaden the definition of Indo-Europeans by including families like his own, which had been present in the Indies for several generations and had acquired social status and economic success. This change in definition in effect diminished associations between Indo-Europeans on the one hand, and poverty, marginality and deprivation on the other.

Koks opened his monograph with the observation that, as a consequence of the almost exclusively male nature of migrants to the Indies for nearly three centuries, European blood had penetrated the colonies along several different avenues: it had gone 'to the *kampung* or to the *Koningsplein* [King's Square, Batavia's most exclusive address dotted with the most impressive mansions; today's *Medan Merdeka*]' (1931: 10). Indo-Europeans could therefore be found at both extremes of the European social spectrum and everywhere in between; consequently, generalizations applicable to all Indo-Europeans were bound to be inadequate. According to Koks, it was necessary to 'differentiate different types of Indo-Europeans, which no author has done thus far' (ibid.: 213). He started out by identifying three groups (ibid.: 13, 19). The first consisted of impoverished Europeans who had taken up residence in the *kampung* and had become virtually indistinguishable in habits and lifestyle from the indigenous population. The second group consisted of soldiers in the colonial army, many of whom had native concubines, a practice condoned by army officers. Upon retirement from the colonial army, many former soldiers joined the first group. The third group consisted of respectable middle-class Indo-Europeans who had made up the greater part of European civil society from the seventeenth century on. The upper layers of this group were not generally viewed as Indo-European at all as they had taken on the characteristics of respectable, upper-middle class *totok* European society. Among members of this group, indigenous ancestry was a topic that, as a matter of good taste, was not considered a suitable topic of conversation – most were more than happy to pass as *totok* European. According to Koks, the status of this group as full-blooded Europeans was a 'very broadly formulated fiction, which was determined by social position' (ibid.: 84). This last group was generally ignored in discussions on the 'Indo problem', giving these debates a somewhat narrow and artificial character. If socially respectable and financially successful individuals were left out of the equation, it could not be surprising that poverty appeared as an inherent characteristic of all Indo-Europeans. Koks argued that an adequate perspective on Indo-Europeans must necessarily include all three groups; discussions would thereby acquire a more nuanced character and the conflation of social and racial categories would be difficult to maintain.

According to Koks, estimates of the number of Indo-Europeans in the colonies only included the group of impoverished Indo-Europeans, defined in purely social terms. After adopting a definition of Indo-Europeans derived from physical anthropology as individuals with mixed indigenous and European ancestry, Koks stated that race-mixing had occurred at a much larger scale than officially acknowledged: 'Indo-Europeans whose descent of *pure* European stock is *very* doubtful, but whose social standing requires it, will be assigned to categories in which they, anthropologically speaking, do not belong' (ibid.: 19). According to the 1930 census, there were around 170,000 individuals of European status who were born in the Indies; the total European population in was 244,000 (these numbers did not include soldiers) (Bosma 1989: 20).[19] It is impossible to estimate the number of individuals of hybrid ancestry who had been classified as native, but their number must be considerable. At this time, it was estimated that Java was inhabited by 42 million people (the census did not include the indigenous population of the Indies).[20] Since the colonial administration did not recognize 'Indo-European' as an official category, the number of Europeans born in the colonies only approximates the number of Indo-Europeans living there. Yet, according to Koks, over 90 per cent of the European inhabitants of the Indies were Indo-European when one follows an anthropological definition and matters of social class are ignored.

Koks was particularly concerned that the common practice of underestimating the number of mixed-race individuals went hand in hand with minimizing their role and accomplishments in colonial society:

> Continuously identifying the accomplishments of Indo-Europeans, in particular if these are exquisite, as those of pure Dutch individuals, while simultaneously assigning an inferior meaning to the concept of 'Indo' with the suffix 'European', has resulted in the predicament that, in the Netherlands, a complete misrepresentation has taken root about the social conditions that obtain in our Indies (1931: 21).

According to Koks, 'practically everything that is reputed to be European in our Indies is, in fact, European *cum grano salis*' (ibid.: 19). He argued that the practice of ascribing all colonial achievements to 'full-blooded' Europeans encouraged a denigrating attitude towards Indo-Europeans and a continued focus on their broken Dutch, their poverty, and their indigenous tastes, manners and lifestyle.

Koks briefly discussed research in physical anthropology, even though he was rather dismissive of it: 'Everybody is sufficiently familiar with the frequent juggling with the concept of race' (ibid.: 102). Most anthropological research in the Indies was conducted on indigenous groups, in particular on those groups which had not experienced much contact with Europeans and, therefore, presumably, with Western civilization. According to Koks, anthropologists had mostly neglected racially hybrid groups even though these were far more interesting (Koks mentioned the work of Eugen Fischer and Ernst Rodenwaldt as exceptions) (ibid.: 101).[21] Koks challenged the assumption that hybrid groups were the result of race mixing and instead argued that it would be more accurate to view them as the result of fusion of two previously separate populations which themselves were already hybrid, pointing out that even the inhabitants of Europe had not been racially pure since the Stone Age (1931: 111).[22] In addition, Koks noted that the physiological measurements collected by anthropologists did not indicate anything about the social life of this interesting in-between group. Even though some anthropologists had argued that mixed-race individuals were better able than Europeans to live, work and procreate in the tropics, they had generally overlooked the fact that mixed-race groups could perish through many other causes than the tropical environment, the incompatibility of inherited features, or degeneration (Koks mentioned unjust and discriminatory colonies policies as far more important factors) (1931: 121). In general, Koks opined, scholars in the social sciences invoked the race concept when no adequate explanations could be found, which led to it becoming 'a commonplace for unsolvable problems' (ibid.: 126). With respect to physical anthropology, Koks stated, 'we can testify that we learned how things should not be done' (ibid.: 129).

Koks noted on several occasions that defining the category 'Indo-European' was unusually difficult because of the flexibility with which it was applied in Indies everyday life. He noted that even the participants at the founding meeting of the Indo-European Association were unable to arrive at a consensus definition; opinions on who was and who was not an Indo-European differed among Indo-Europeans as well (ibid.: 131). During the 1910s, a number of Indo-European groups had admitted educated Indonesians as members because they considered them respectable and fully Westernized individuals who were, in many respects, similar to Indo-Europeans. Similarly, some of these groups also admitted the offspring of *totok* Europeans who were born in the Indies and whose parents had not been able to afford to send them to the Netherlands for their education – socially, these individuals encountered the same disadvantages as Indo-Europeans but they did not have any indigenous ancestry. Yet considering educated Indonesians and creole *totok* Europeans as Indo-Europeans remained controversial: neither group had a mixed European and indigenous ancestry even though both probably shared several social characteristics of Indo-Europeans.

Koks presented a definition of Indo-Europeans as individuals who had both European and Indonesian ancestry – irrespective of how many generations ago a mixed marriage might have taken place. He ended with a much broader and more inclusive definition, which followed indigenous Indonesian conceptions of identity, in which one's place of birth was of paramount importance. An Indo-European, in Koks' definition, is a European who was born in the Indies: 'Indo is he, who the natives call *peranakan Blanda* [Malay: a European born in the Indies], born in this country, *bangsa Blanda* [Malay: European nation/tribe/group]. It is someone whose father is European *de jure* [who is legally classified as European]' (ibid.: 136). Following this definition, Koks himself was unquestionably Indo-European. Defined this way, the great majority of Europeans residing in the Indies, including full-blood Europeans without any indigenous ancestry, were defined as Indo-Europeans. Considerations of Indo-Europeans which focused merely on the poor, marginalized and down-trodden Europeans were therefore overly narrow and misguided. Moreover, Koks argued that a natural psychic bond tied Indo-Europeans and Indonesians together; being born in the Indonesian archipelago connected both groups and made them often feel an aversion to *totok* Europeans (the so-called migrants) who aimed to avoid contamination by anything from the Indies.

One issue which inspired official concern about indigent Indo-Europeans and fuelled public discussions on the Indo problem was the issue of sexual impropriety. Many poor Europeans (and Indo-Europeans) shared their lives in the *kampung* with an indigenous woman without being officially married.[23] It was not uncommon for recently arrived Europeans to acquire a concubine (euphemistically called a housekeeper or, in derogatory terms, a piece of furniture [in Dutch: *meubelstuk*]). When they had acquired sufficient wealth, they would discard their concubine and marry a European woman. At times, these men adopted their mixed-race children (named pre-children [in Dutch: *voorkinderen*]) or had them adopted by a friend or family member; if no action was undertaken, these children would be classified as native. The nature of these illegitimate relationships was of great concern to upstanding and respectable Europeans in the Indies. As Koks argued, whether these relationships were legitimate or not was not the main issue: 'We see many illegitimate relationships there [in the Indies], from which many legal relationships could learn a few things' (1931: 217). Many of these illegitimate children found a way to acquire European legal status by convincing a European to adopt them, at a 'tariff, generally, of a "square jar", a square bottle of *jenever* [Dutch gin]' (ibid.: 144). Such practices were generally considered highly scandalous, indicating the low-life morals and utter depravity of impoverished Europeans. Yet Koks described this transaction as a welcome alternative for individuals of European ancestry who had become outlawed because of the accidental circumstances around their birth and the inflexibility of colonial legal provisions.

In addition to providing an alternate account of Indo-Europeans in the Dutch East Indies, Koks is at his best when he undermines preconceived ideas and theoretical accounts of them. He uses many pointed anecdotes to illustrate his arguments, to undermine legal and racial categories used in the Indies, and common views on Indo-Europeans in colonial life. At one point, he relates how a physician instantly fired his native driver after discovering that he was the son of a wealthy Italian marble sculptor (and therefore an Indo-European). Apparently, this physician found it improper to hire an Indo-European for this demeaning job – unwittingly reducing his former employee to poverty. In a similar vein is the anecdote of a man who applied for a position of native chauffeur armed with a letter of recommendation written by a professor – who, as it turned out, was his own father (who, probably because of embarrassment, had not acknowledged his son).[24] In a third anecdote, Koks related an uncomfortable moment when a high government official exclaimed that he detested Indo-Europeans. An upper-echelon bureaucrat retorted by mentioning that he, himself, was Indo-European. The official, clearly discomfited, replied that, even though this government employee had very dark facial features, he surely was not an Indo, probably because he thought that being an Indo-European was not compatible with the bureaucrat's high social standing (1931: 28). In a final anecdote,

Koks related his experiences as a member of the board of an Indo-European group in the Netherlands, where a rather peculiar method of approving applicants for membership was adhered to: 'Although this association even counted among its members a few individuals who had never been to the Indies, but only if they had a character that corresponded to that of a respectable Indo, it never hesitated to exclude those individuals who were Indo-European by birth but who were part of the less respectable Indo population' (ibid.: 131). In these anecdotes, Koks was able to convey the reality of everyday life of Indo-Europeans in unusual clarity.

In a humorous way, Koks undermined and deconstructed several statements on Indo-Europeans made by a range of experts. He quotes, for example, a book on the city Batavia, which mentioned that the 'group of Europeans, also including the Indo-Europeans, is about 30,000 individuals strong' (1931: 20).[25] In a similar vein, Koks argues, one could state about Amsterdam that 'the group of Jews, also including the adherents of the New Testament, is about 750,000 people strong' (ibid.). In other words, such statements were utterly meaningless. By interjecting critical notes in extensive quotes by experts, Koks criticized them while citing their views. By quoting social democrat politician J.E. Stokvis extensively, for example, Koks provided a running commentary by interjecting his critique within brackets. Stokvis, for instance, argued that 'the mental life of the Indo-European is more inclined to the psyche of the indigenous mother (note: this is not always a native woman) than to that of the white father (in ninety out of a hundred cases an Indo-European…)' (ibid.: 108). In this way, Koks conveyed that most Indo-Europeans had Indo-European parents. He also poked fun at relatively common observations, such as the comment made by many first-time visitors to the Indies who noted that so many Indo-Europeans sported typical 'Indies' faces. Koks responded that a single visit to the Netherlands would reveal that 'these so-called Indies faces are found in great numbers among Dutch people who have never mixed with coloured ones' (ibid.: 119). He took aim at the grandiose attempts of anthropologists to describe the superiority of the white race. 'Imagine a Chinese or African anthropologist,' he stated, 'who explains the crisis [the Great Depression] by referring to the nature of the white race!' (ibid.: 127). According to Asian anthropologists, he stated, 'all European thinkers would have been Mandarins' (ibid.: 168).

Concluding Remarks

Apart from presenting a cogent definition of 'Indo-European', Koks excels when presenting puns, anecdotes and riddles undermining commonly accepted views on race and racial difference in the Indies. He presents his views in a rather meandering way, which makes it at times relatively hard to follow his arguments. In his dissertation and subsequent book on Indo-Europeans in the Dutch East Indies, Koks aimed to reframe the debate on Indo-Europeans by diminishing the emphasis on stigmatized groups consisting of impoverished Europeans maintaining a lifestyle which was considered immoral, depraved and un-European, and by including the much larger group of upstanding and respectable citizens who had been present in the Indies from the seventeenth century on and who were generally considered European because of their social standing. By emphasizing ancestry and place of birth, instead of then-current social definitions of what it meant to be European, he broadened existing discussions and aimed to diminish the tone of moral reprobation that characterized them. In Koks' more encompassing definition, Indo-Europeans were responsible for almost everything Europeans had accomplished in the colonies – they therefore are more adequately described as pioneers rather than paupers.

Subsequently, Koks provided sociological explanations for the social and economic position of different sub-groups of Indo-Europeans by referring to factors including lack of educational opportunities, the discriminatory attitude of *totok* Europeans, the competition with educated Indonesians, and the prohibition to own land. These analyses contain few surprises for today's readers but were novel at the time. Koks illustrated the artificial nature of current definitions of Indo-Europeans, underscoring the fact that the category did not even exist officially, which doomed any attempt to collect statistical data or any other form of information from the outset. By using various examples and anecdotes, and by illustrating the often absurd simplicity of current discussions on the Indo problem, be they informal, bureaucratic or scholarly, Koks undermined and deconstructed generally accepted views on Indo-Europeans in the Dutch East Indies. As Stoler has argued, Indo-Europeans constituted a creative and disruptive force that tried to shift colonial boundaries of racial exclusion. Koks' writings have taken on this same quality as a creative and disruptive force undermining and deconstructing artificial exclusionary boundaries in colonial life. This could be the main reason that his book was ignored in Dutch academic circles and in intellectual debates in the Dutch East Indies. Even the magazine of the Indo-European Association hardly mentioned it.

In the Dutch East Indies, individuals defined as Indo-Europeans found themselves in a paradoxical situation: even though they were members of a group that did not officially exist, they were the most extensively investigated group in the colonies. Yet the several commissions investigating poverty among Europeans in the Indies did not aim to analyse human variation; instead, they tended to homogenize Indo-Europeans as a group halfway between Europeans and Indonesians and precluded recognizing and interpreting variations within this group. The focus of official attention was structured using artificial legal categories that did not map onto racial or ethnic ones. The attention to race was often paired with an interest in prurient sexuality, as Indo-Europeans were considered to be the fruit of transgressive relationships between Europeans and Indonesians, and thereby found its origin in a transgression of *totok* European civic morality. In lifestyle and appearance, this group violated expectations of what it meant to be European in the colonies. In the colonial imagination, the condition of this group was interpreted as a racial issue in terms of being an undesirable and objectionable admixture with the indigenous population through weakness of will, lack of morality, mistaken opportunism or degeneration. In contrast to the homogenizing effects of these official investigations and broader public discussions on the Indo problem, Koks introduced a perspective on Indo-

Europeans which emphasized variability and, with it, the porousness, arbitrariness, and at times the absurdity of colonial categories of racial classification.

Koks' writings illustrate the flexibility and malleability of racial thinking, as he was able to employ existing bureaucratic categories imaginatively to come to completely different conclusions about the position and role of Indo-Europeans in the Dutch East Indies. Yet, at the same time, his work demonstrates the intransigence of racial categorization in the Dutch East Indies. As colonial life was based on the legal distinction between Europeans, foreign Orientals and natives, individuals or groups who did not easily fit in these categories would either be placed, more or less arbitrarily, in one of them or be ignored altogether. As a consequence, the statistical data collected by the colonial administration through census, taxation records and other means did not allow it to investigate the social position of Indo-Europeans, their type of employment, their average income, or anything else for that matter. Specific contingencies of colonial categorization and hence bureaucratic documentary practices in the Dutch East Indies constituted the condition of possibility for knowledge of human variation; however, in this case, they precluded gaining any insight into Indo-Europeans. Even though officials and administrators often governed by acquiring data on their subject populations, at times they saw greater advantage in not knowing about conditions among specific population groups. Recently, the social construction of ignorance, as an activity equally important as the social construction of knowledge, has received attention from scholars in the field of science studies.[26] The intransigence of bureaucratic classifications in combination with the persistence of racial stereotypes towards Indo-Europeans in the Dutch East Indies can explain the resilience of the latter, as no data was available to correct them. At the same time, negative views about Indo-Europeans could easily be confirmed on an anecdotal basis. In other words, everyday colonial administrative practices, which could not provide any information about Indo-Europeans, existed alongside broader social ideas on poor, deceptive, lazy and parasitic Indo-Europeans. What the statistical data at their disposal did not tell them they knew already because they shared these social views. On the occasions when they were called to address inequities in income, educational and vocational opportunities, health and longevity, colonial administrators could easily resist undertaking any action by relying on the slippage between legal, anthropological and biological categories, which allowed them to conclude that the problem they were asked to address did not exist.

During colonial times, Koks did not publicly express strong opinions on the Indonesian nationalist movement except for pointing out that, generally, such movements often developed in close relationship with political ferment among mixed-race individuals. He mentioned several associations and political parties in which Indo-Europeans and Indonesians cooperated.[27] During the Japanese occupation, Koks was interned like most European men who had been called up to serve in the colonial army. After the declaration of Indonesian independence on 17 August 1945 and the transfer of sovereignty on 27 December 1949, Koks enthusiastically supported the Republican cause and became a professor of Economics at the University of Hasanuddin in Makassar (Sulawesi) until 1962 (Makaliwe 1963:

513).[28] His conviction that human communities should be defined in terms of psychic bonds and shared culture rather than in terms of race, in particular when those definitions support the social attitudes of specific groups, combined with his view that European, Indo-European and Indonesian culture had deeply penetrated each other for over three hundred years, led him to embrace the new state. He regularly wrote snarky opinion pieces in the Dutch-language paper *De Nieuwsgier*. He died in Jakarta in the early 1970s.[29]

Hans Pols is associate professor at the Unit for History and Philosophy of Science at the University of Sydney. He is interested in the history of medicine, in particular the history of psychiatry. He has conducted research on the history of psychiatry in the United States, the conceptualization of war neuroses during the Second World War, the history of colonial psychiatry in Indonesia, and the history of the Indonesian medical profession. He is currently engaged in a research project on the recent past, the present condition, and the future of psychiatry in Indonesia.

Notes

1. See Kielstra (1929) and de Grave (1938).
2. The dissertation appeared with a rather bland title: J.T. Koks. 1930. *Koloniaal-Sociografische Bijdrage*. Amsterdam: H.J. Paris.
3. In this chapter, I will use the phrase 'native' even though this has become unfashionable as of late because of its derogatory connotations. I will continue to use this phrase for two reasons: first, I do not wish to represent colonial discourse in a more favourable light than it really was; second, when I use it, it refers to the legal category rather than to specific individuals or populations.
4. For an overview of the legal categories in the Dutch East Indies and what was involved in applications to be considered 'equivalent to European', see Luttikhuis (2013).
5. See also Bosma and Raben (2008).
6. On the role of European women in maintaining an exclusive European identity in the Indies, see Stoler (2002a), Locher-Scholten (2000) and Clancy-Smith and Gouda (1998).
7. Ann Stoler's work has pioneered new analytical approaches to the discourse on Indo-Europeans in the Dutch East Indies. See, for example, Stoler (2002a).
8. For an insightful discussion on how categories of race, social class and social status were related to each other and often defined in terms of each other, see Protschky (2011).
9. For an overview of the history of the Indo-European Association, see Bosma (1997).
10. For more on Veth, see Beekman (1986).
11. For theories about the adverse influence of the tropical climate on the health of Europeans, see Pols (2012), Livingstone (1999) and Anderson (2006).
12. For an analysis of literary depictions of spirits, incantations and the like among Indo-Europeans and the indigenous populations of the Indies, in particular how they could affect romantic relationships between European men, and Indo-European and indigenous women, see Wiener (2007).
13. On Indonesian concubines, see Baay (2008).
14. As examples can serve Couperus (1900) and Maurits (1893).

15. For an overview of Dutch colonial literature, with many observations like these, see Nieuwenhuys (1982).
16. The biographical information on J.T. Koks and his family is derived from the annual list of European inhabitants published by the colonial administration of the Dutch East Indies and family notices that appeared in Dutch East Indies newspapers as accessed through http://www.delpher.nl.
17. Koks' wife was born in Medan (Sumatra); her father was Czech and her mother was Indo-European. The tobacco and rubber plantations surrounding Medan, the urban centre of Deli on the East Coast of Sumatra, attracted men from all over the world who were eager to make their fortune. Conditions in Deli have been described in several colonial novels; see, for example, Székely-Lulofs (1933) and Stoler (1985).
18. Koks' dissertation was entitled *Koloniaal-Sociografische Bijdrage* (Colonial Sociographic Contribution).
19. Using the same data of the 1930 census, Paul van der Veur estimates the number of Indo-Europeans to be 134,000, or 56% of all Europeans living in the Dutch East Indies (1954: 125).
20. These figures are based on the 1930 census. See The Population of Indonesia (1974: 8).
21. For Dutch research on physical anthropology, see Sysling (2013). Eugen Fischer (1913) wrote one of the first extensive studies on race mixing. Ernst Rodenwaldt (1927) wrote an elaborate, two-volume study on a mixed-race group which lived on a small island near East Timor.
22. This was a common view among human biologists in the Netherlands. This view was presented in many places; see, for example, van Herwerden (1926). Van Herwerden was the most outspoken advocate of eugenics in the Netherlands.
23. It is of course Ann Stoler who has drawn our attention to the moral concerns about these relationships, see Stoler (2002b).
24. Both anecdotes are given in Koks (1931: 17, footnote 1).
25. The book on Batavia Koks cited is van der Zee and Meyroos (2005).
26. See, for example, Proctor and Schiebinger (2008).
27. In his work, Ulbe Bosma had analysed the close ties between Indonesian nationalism and increasing Indo-European consciousness. See Bosma (1997, 2004).
28. For his writings after 1950, see Koks (1951a, 1951b).
29. Personal communication from Lana Frost, grand-daughter of J.T. Koks, to the author, 20 December 2014.

Bibliography

The biographical information on J.T. Koks and his family is derived from the annual list of European inhabitants published by the colonial administration of the Dutch East Indies (*Regeeringsalmanak voor Nederlandsch-Indië and Nieuw Adresboek van Geheel Nederlandsch-Indië*) and family notices that appeared in Dutch East Indies newspapers as accessed through http://www.delpher.nl.

Alatas, H.S. 1977. *The Myth of the Lazy Native: A Study of the Image of the Malays, Filipinos and Javanese from the 16th to the 20th Century and Its Function in the Ideology of Colonial Capitalism*. London: F. Cass.

Anderson, W. 2006. *Colonial Pathologies: American Tropical Medicine, Race, and Hygiene in the Philippines*. Durham: Duke University Press.

Baay, R. 2008. *De Njai: Het Concubinaat in Nederlands-Indië*. Amsterdam: Athenaeum-Polak & Van Gennep.

Beekman, E.M. 1986. 'Bas Veth: A Colonial Muckraker', *Indonesia* 42: 101–112.

Birney, A. 2012. *De Dubieuzen: Wat Multatuli, Daum en Couperus Ons Niet Vertelden (en Wij Niet Konden Lezen)*. Haarlem: In de Knipscheer.

Bosma, U. 1989. 'De Indo-Europeaan en de Autonomie voor Indië', *BMGN: Low Countries Historical Review* 104(1): 17–38.

_____ 1997. *Karel Zaalberg: Journalist en Strijder voor de Indo*. Leiden: KITLV Press.

_____ 2004. 'Citizens of Empire: Some Comparative Observations on the Evolution of Creole Nationalism in Colonial Indonesia', *Comparative Studies in Society and History* 46(4): 646–681.

_____ 2005. 'The Indo: Class, Citizenship and Politics in Late Colonial Society', in J. Coté and L. Westerbeek (eds), *Recalling the Indies: Colonial Culture & Postmodern Identities*. Amsterdam: Aksant, pp. 67–97.

Bosma, U. and R. Raben. 2008. *Being 'Dutch' in the Indies: A History of Creolisation and Empire, 1500-1920*, trans. W. Shaffer. Singapore: NUS Press.

Clancy-Smith, J. and F. Gouda (eds). 1998. *Domesticating the Empire: Race, Gender, and Family Life in French and Dutch Colonialism*. Charlottesville: University of Virginia Press.

Couperus, L. 1900. *De Stille Kracht*. Amsterdam: L.J. Veen.

_____ 1985. *The Hidden Force*, translated by E.M. Beekman. Amherst: University of Massachusetts Press.

De Grave, F. 1938. *Het Indo Probleem*. Semarang: Bruin.

Fasseur, C. 1994. 'Corner Stone or Stumbling Block: Racial Classification and the Late Colonial State in Indonesia', in R.Cribb (ed.), *The Late Colonial State in Indonesia: Political and Economic Foundations of the Netherlands East Indies, 1880-1942*. Leiden: KITLV Press, pp. 31–56.

Fischer, E. 1913. *Die Rehobother Bastards und das Bastardierungsproblem beim Menschen: Anthropologische und Ethnographische Studien am Rehobother Bastardvolk in Deutsch-Südwest-Afrika*. Jena: Gustav Fischer Verlag.

Kielstra, J.C. 1929. 'The "Indo-European" Problem in the Dutch East Indies', *Asiatic Review* 25: 588–595.

Koks, M.T. 1922. *Het Arabinezuur als Vector voor de Bereiding van Geneesmiddelen*. Amsterdam: Johannes Jesse.

Koks, J.T. 1930. *Koloniaal-Sociografische Bijdrage*. Amsterdam: H.J. Paris.

_____ 1931. *De Indo*. Amsterdam: H.J. Paris.

_____ 1951a. *Nationalisme en Nationale Economie*. Bandoeng: Quick.

_____ 1951b. *Endjeniring Tentara*. Djakarta: Penerbitan dan Balai Buku Indonesia.

Livingstone, D.N. 1999. 'Tropical Climate and Moral Hygiene: The Anatomy of a Victorian Debate', *British Journal for the History of Science* 32(1): 93–110.

Locher-Scholten, E. 2000. *Women and the Colonial State: Essays on Gender and Modernity in the Netherlands Indies, 1900-1942*. Amsterdam: Amsterdam University Press.

Luttikhuis, B. 2013. 'Beyond Race: Constructions of "Europeanness" in Late-Colonial Legal Practice in the Dutch East Indies', *European Review of History* 20(4): 539–558.

Makaliwe, D.H. 1963. 'Vijfteen Jaren Economisch Hoger Onderwijs te Makassar', *De Economist* 111(7/8): 510–520.

Maurits [P. A. Daum]. 1893. *'Ups' en 'Downs' in het Indische Leven*. Batavia: Kolff.

Monnais, L. and H. Pols. 2014. 'Health and Disease in the Colonies: Medicine in the Age of Empire', in R. Aldrich and K. McKenzie (eds), *The Routledge History of Western Empires*. New York: Routledge, pp. 270–284.

Nieuwenhuys, R. 1982. *Mirror of the Indies: A History of Dutch Colonial Literature*, trans. F. van Rosevelt. Amherst: University of Massachusetts Press.

Pols, H. 2010. 'Eugenics in the Netherlands and the Dutch East Indies', in A. Bashford and P. Levine (eds), *The Oxford Handbook of the History of Eugenics*. London/New York: Oxford University Press, pp. 347–362.

——— 2012. 'Notes from Batavia, the European's Graveyard: The 19th Century Debate on Acclimatization in the Dutch East Indies', *Journal of the History of Medicine and Allied Sciences* 67(1): 120–148.

The Population of Indonesia. 1974. *The Population of Indonesia*. Jakarta: Lembaga Demografi, Universitas Indonesia.

Proctor, R. and L. Schiebinger (eds). 2008. *Agnatology: The Making and Unmaking of Ignorance*. Stanford: Stanford University Press.

Protschky, S. 2011. 'Race, Class, and Gender: Debates over the Character of Social Hierarchies in the Netherlands Indies, Circa 1600–1942', *Bijdragen tot de Taal-, Land- en Volkenkunde [BKI]* 167(4): 543–556.

Rodenwaldt, E. 1927. *Die Mestizen auf Kisar*. Jena: Verlag von Gustav Fischer.

Stoler, A.L. 1985. *Capitalism and Confrontation in Sumatra's Plantation Belt, 1870-1979*. Ann Arbor: University of Michigan Press.

——— 2002a. *Carnal Knowledge and Imperial Power: Race and the Intimate in Colonial Rule*. Berkeley: University of California Press.

——— 2002b. 'Sexual Affronts and Racial Frontiers: Cultural Competence and the Dangers of Métissage', in *Carnal Knowledge and Imperial Power: Race and the Intimate in Colonial Rule*. Berkeley: University of California Press, pp. 79–111.

——— 2009. *Along the Archival Grain: Epistemic Anxieties and Colonial Common Sense*. Princeton: Princeton University Press.

Sysling, F. 2013. 'Geographies of Differrence: Dutch Physical Anthropology in the Colonies and the Netherlands, Ca 1900-1940', *BMGN: Low Countries Historical Review* 128(1): 105–126.

Székely-Lulofs, M. 1933. *Rubber: A Romance of the Dutch East Indies*, trans. G.J. Renier and I. Clephane. London: Cassell.

Taylor, J.G. 1983. *The Social World of Batavia: European and Eurasian in Dutch Asia*. Madison: University of Wisconsin Press.

Van der Veur, P.W. 1954. 'The Eurasians of Indonesia: Castaways of Colonialism', *Pacific Affairs* 27(2): 124–137.

Van der Zee, D. and A. Meyroos. 2005. *Batavia: De Koningin van het Oosten*. Rotterdam: Schueler.

Van Herwerden, M.A. 1926. *Erfelijkheid bij den Mensch en Eugenetiek*. Amsterdam: Nederlandsche Bibliotheek.

Veth, B. 1900. *Het Leven in Nederlandsch-Indië*. Amsterdam: Van Kampen.

Widmer, A. 2012. 'Of Field Encounters and Metropolitan Debates: Research and the Making and Meaning of the Melanesian "Race" during Demographic Decline', *Paideuma* 58: 69–93.

Wiener, M. J. 2007. 'Dangerous Liaisons and Other Tales from the Twilight Zone: Sex, Race, and Sorcery in Colonial Java', *Comparative Studies in Society and History* 49(3): 495–526.

AFTERWORD
Following Racial Paper Trails

Warwick Anderson

Racial thought in the twentieth century – whether in physical anthropology, human biology or clinical science – was accommodated in files and registers. Millions of people were made into scientific facts and mobilized in one such form or the other. Thus they entered the modern 'scriptural economy', as Michel de Certeau (1984: 131) would put it, and became visible and calculable – or reified. But what do we know of these 'literary technologies' (Shapin and Schaffer 1984) that turned human difference into race?[1] How, as Alexandra Widmer and Veronika Lipphardt ask in this volume, might we delineate and trace the paper trails of racial thought in the twentieth century? What, in particular, were the blueprints and specifications of the documentary machinery of the colonial state and its racializing agents? These are pressing questions, especially as new electronic databases offer to substitute for the file and the register in the framing of race in the twenty-first century. What form, we may wonder, will 'paperless' race take? What relation will it claim to the racial figures that emerged from older writing technologies?

According to Bruno Latour, the sure path of science follows 'the construction of well-kept files in institutions that want to mobilize a larger number of resources on a larger scale' (1986: 15).[2] Latour was echoing Michael Foucault, who earlier had inquired into the problem 'of the entry of the individual description, of the cross-examination, of anamnesis, of the "file" into the general functioning of scientific discourse' (1979: 191) during the eighteenth century. But Latour would be both more forensic and more expansive in his examination of the ways in which the modern file or dossier mobilized facts. He wanted to see how scientists made people, for example, into immutable, presentable, readable and combinable things that might then be mustered or circulated. Paperwork was crucial. 'Mobilization is not restricted to paper', he wrote, 'but paper always appears at the end when the scale of this mobilization is to be increased'. As scientists accumulated 'traces' and simplified inscriptions, they produced harder facts. Once abstracted, these resources could be further mobilized, or scaled up and down, through more paperwork. Therefore 'by working on papers alone, on fragile inscriptions which are immensely less than the things from which they are extracted, it is still possible to dominate all things, and all people' (1986: 17, 32). Later, Latour would call for a 'meticulous investigation into dossiers – grey, beige, or yellow; fat or thin; simple or complicated; old or new – to see where they lead us' (2009: 70). Surely it is timely, now, to ask where such documents lead us in the sciences of race?

Nearly a century ago, Max Weber touched on the significance of files, dossiers and other paperwork in modern bureaucracy – but he did not tarry (1978 [1922]). Since then we have learned much more about the contributions of office technologies, especially since the late nineteenth century, to bolstering and extending bureaucratic rationality, to making and distributing facts. JoAnne Yates (1989) has drawn attention to typewriters and their popularity after the 1880s; the fin-de-siècle uptake of carbon paper and stencil duplication; the propagation of ring binders; and perhaps most pertinently, the development of vertical filing, which proliferated in the early twentieth century. From the 1890s, card indexes and file folders, often made from Manila hemp, could be arranged vertically and ordered in the new filing cabinets.[3] Several historians have followed the paper trail of industry and the state, the office machine 'revolution', further into the twentieth century. Jon Agar (2003) described British government efforts to assemble central registers after the First World War, in order to identify and inspect national subjects.[4] Major American and European hospitals in the 1920s began reforming their record keeping along military and criminological lines, creating individual case files for patients, a 'unit medical record' that tracked admission courses and illness careers (Anderson 2013). In the modern inter-war economy, experts sought efficiency, system, indexicality and seriality in documentation, though not always mobility, since durability usually seemed more important than portability. As the volume of data grew, and management theories took hold, further innovation in information processing occurred. Punched card data-processing using Hollerith machines, or mechanical tabulators, first rendered more efficient the counting of the United States census in 1890. After the 1920s, such mechanical data processing became widespread (Agar 2003; Müller 2014). In Germany, punched cards and Hollerith machines enabled the identification, registration and surveillance of the whole population – in a sense, they made 'population' visible – and thus formed a necessary component of National Socialist information infrastructure (Aly and Roth 2004 [1984] and more provocatively Black 2001).[5] How, we should ask again, did these new technologies for recording, documenting and ordering facts about people reshape racial registration in science during the twentieth century – in other words, how might writing machines like the typewriter make types, or tabulators punch out populations?

It is important to emphasize that different paper trails tell different stories about us. We need therefore to consider the specific qualities and potentials of the paper technologies that conjure up race and population, among other things. Or as Matthew S. Hull (2012a) put it, we should treat documents as 'mediators that shape the significance of the signs inscribed on them and their relations with the objects [or persons] they refer to' –

that is, the materiality of paperwork matters. Foucault and others have argued that documents – whether files, dossiers, reports or registers – generate different kinds of authors, subjects, populations and readers. They make visible different things and different collectives. This does not necessarily imply that they always function as mechanisms of control and discipline. Indeed, most of those subject to bureaucratic rationality, to that particular scriptural economy, escape its operations, or remain indifferent to them. Nonetheless, practices of documentation do shape patterns of scientific knowledge – often in ways we are yet to understand fully. They establish a range of parameters, rules and protocols for explaining human difference; they have disparate informatic sensibilities; they show, according to Ben Kafka, distinct dimensions of attachment or aggression (2012: 108). The individual file, for example, might reveal heteroglossia, or at least a dialogic process, in its formation. Often it possesses a provisional, interactive character, 'marginalia' in which may be glimpsed a connection to the person or composite represented, a spectral presence – or perhaps just a faded, awkward photograph affixed to a corner. In contrast, the registry, like the punched card, can serve to sunder such relations. 'Lists do not communicate', Cornelia Vismann has claimed, 'they control transfer operations' (2008: 6). That is, they mobilize, and regulate the passage of, people and goods. In her view, the registry is a 'monologue of power that installs itself as absolute, a mechanism of colonial or state sovereignty' (ibid.: 95).[6] So what was happening, then, when race scientists set aside their note books, index cards and files, which had generated Mendelian types and racial composites, and took up punched cards and tabulators, which eventually registered evolutionary populations? What quality of attachment or aggression might this transition, perhaps paradoxically, suggest to us?

We still need to learn much more about the paperwork of race science in the twentieth century. This volume moves us along the path. In the introduction, Widmer and Lipphardt focus their analysis on the documentary machinery of colonial race science, on the various papers that clutter our contemporary anthropological archives. They concentrate on the representational aspects of all this paperwork, its consistent concealment or erasure of close encounters in the field – in effect, distinguishing the field from the filed. At the same time, they are exceptionally sensitive to the 'material, technical, and processual lives' of these literary technologies and retrieval systems, showing the entanglement of scientific and other colonial bureaucratic practices, all of them dedicated to identifying and controlling human difference.[7] The contributors to this volume in their chapters adduce an eclectic array of literary technologies, including field notes, index cards, files, dossiers, registers, censuses, correspondence, reports, articles and books, as though their anthropologists, scientists and medical officers had been feverishly committed to papering over the world. This global paper shuffling reveals multiple bureaucratic logics in race science. Samuël Coghe, for example, reflects on how registration cards and register books produce a racialized population different from the individualized records of oral histories and interviews (this volume).[8] Toward the end of this period, many physical anthropologists and human biologists, seeking to encompass larger and larger populations, would turn also to data processing. In the late 1920s, the graduate students of Earnest Hooton at Harvard began to use punched cards and Hollerith machines to tabulate mixed-race populations from around the world, sending their burgeoning data to the 'Peabody girls' to be processed – the basement-dwelling female technicians who would come to rival 'native' intermediaries in their invisibility in the race science writing machine (Shapin 1989).[9] Anthropologists seeking to introduce 'population thinking' into race science seized the means of its production in the 1930s and 1940s, rendering their studies more rigorously and aggressively 'scientific' and objectifying. Most of them forsook case studies, individual files and dossiers – which often had displayed a rather discomforting libidinal investment in the exotic or primitive, a lingering romantic attachment beneath outward disparagement – in favour of data processing and mechanistic calculation, which generated, or at least made visible, populations ripe for development and modernization in the coming Cold War. Thus following the paper trail can be another way of following the money.

Paperwork inevitably presents problems of coordination and standardization, or metrology. How do we compare observations and measurements recorded by different people in different places at different times? This problem is as troubling for those compiling sociological records and files as it is for those who measure bodies or examine blood and tissue samples. In this volume, Maria Letícia Galluzzi Bizzo provides a vivid example of the difficulties in standardizing nutritional assessments at various sites, which called for innovative strategies and improvisations to scale up or down the measuring techniques. Over the past few centuries, standards and classification have proliferated in order to control and stabilize data, as part of new bureaucratic and research infrastructures. To an extent, this sort of categorical activity has acted to discipline researchers and the authors of paperwork. One of the recurrent themes in the archives of North American physical anthropologists in the early twentieth century is the perceived need to standardize anthropometric techniques: whenever they met with colleagues, Charles B. Davenport, Ales Hrdlicka, Clark Wissler and Harry L. Shapiro would meticulously measure one another's bodies and compare results. Additionally, they sought to devise standard forms for anthropometry.[10] These efforts to standardize the measurement and appraisal of bodies were occurring in parallel (if separately) with attempts to make uniform and comparable the diagnosis of disease, to develop the International Classification of Diseases (Bowker and Star 2000).[11] Of course, these endeavours ultimately failed, like all efforts to stabilize knowledge permanently. Moreover, we should bear in mind that the 'economy and ecology of standards' in the twentieth century, even as it strived to make visible and stabilize kinds of persons and diseases, also inevitably rendered invisible whoever and whatever did not fit the regular or conventional categories (Star and Lampland 2009: 7). Thus Hans Pols in his chapter show us how mixed-race people in the Dutch East Indies could disappear if they failed to fit the colonial state's racial categories and filing system.[12] Although 'listening to infrastructure' can be as exciting as listening to a metronome, we need, as historians, to put up with such boredom, for standards, however tedious, are the ways in which bureaucracies 'set parameters within which social action takes place', and they 'constrain a phenomenon within a particular set of dimensions' (Star and Lampland 2009: 13, 14). According to sociologist Lawrence Busch, 'standards are means by which we perform the world' (2011: 13).[13]

Until recently, the material cultures of the sciences of race have escaped the critical review and appraisal given to the literary technologies of cultural and social anthropology.[14] Still more surprising, their materiality also has evaded the sort of exacting analysis to which other parts of the colonial writing machine have been subject.[15] Perhaps this peculiar exemption, this failure to recognize the scriptural economy and standard setting of race science in the twentieth century, reflects the success of its practitioners in abstracting, objectifying and sublimating the contents of research – so much so that form and style can appear trivial and inconsequential. And yet, without their registers and files, their index cards and punched cards, their standard forms and measurements, scientists could never have succeeded in making up races and populations in the ways they did.[16]

Notes

I would like to thank Veronika Lipphardt and Sandra Widmer for their encouragement and assistance in the writing of this chapter. Rachel Ankeny, Adele Clarke, Sebastián Gil-Riaño, Rebecca Lemov, Hans Pols, Joanna Radin, Craig Robertson and Ricardo Roque commented helpfully on earlier versions. A grant from the Australian Research Council (FL 110100243) supported this project.

1. See also Shapin (1984) and the qualifications adumbrated in Schaffer (1998).
2. On the importance of laboratory 'inscription devices', see Latour and Woolgar (1979). It is interesting that 'inscription devices' and 'literary technologies' should come to fascinate scholars in science studies just as writing and paper were about to disappear from most scientific work.
3. See also Rotella (1981), Flanzreich (1993), Dery (1998), Day (2001), Gardey (2008), Krajewski (2011) and Robertson (2015).
4. As Agar points out, 'nearly every historian forgets about the form of the file and reads the content' (2003: 1).
5. On older forms of state identification, see Torpey (2000), Caplan and Torpey (2001) and Robertson (2010).
6. See also Gitelman (2014) and Riles (2006).
7. See also Widmer (2008).
8. See the other chapters in this collection for additional examples.
9. The data processors and stenographers worked in the Peabody Museum at Harvard, hence the 'Peabody girls'.
10. This is based on my observations of the relevant archives. See also Morris-Reich (2013) and Lipphardt and Sommer (2015).
11. See also Thévenot (1984, 2009), Clarke and Fujimura (1992), and Slaton and Abbate (2001).
12. See also Stoler (2009).
13. Busch notes: 'As with all standardizing projects, the colonial project never achieved its goals' (2011: 98).
14. For example, see Clifford and Marcus (1986), Riles (2006), Starn (2015) and Lemov (2016). Lemov's account of 'endangered data' applies more generally to human biology too, of course.

15. On colonial information gathering, see Appadurai (1993), Cohn (1996) and Bayly (1996). For a deeper engagement with colonial literary technologies, see Hawkins (2002), Ogburn (2007), Mueggler (2011) and Hull (2012b).
16. My phrasing deliberately echoes Hacking (1986).

Bibliography

Agar, J. 2003. *The Government Machine: A Revolutionary History of the Computer*. Cambridge MA: MIT Press.

Aly, G. and K.H. Roth. 2004 [1984]. *The Nazi Census: Identification and Control in the Third Reich*. Philadelphia: Temple University Press.

Anderson, W. 2013. 'The Case of the Archive', *Critical Inquiry* 39: 532–547.

Appadurai, A. 1993. 'Number in the Colonial Imagination', in C.A. Breckenridge and P. van der Veer (eds), *Orientalism and the Post-Colonial Predicament*. Philadelphia: University of Pennsylvania Press, pp. 314–339.

Bayly, C. 1996. *Empire and Information: Intelligence Gathering and Social Communication in India, 1780-1870*. Cambridge: Cambridge University Press.

Black, E. 2001. *IBM and the Holocaust: The Strategic Alliance Between Nazi Germany and America's Most Powerful Corporation*. New York: Crown Publishers.

Bowker, G.C. and S.L. Star. 2000. *Sorting Things Out: Classification and Its Consequences*. Cambridge MA: MIT Press.

Busch, L. 2011. *Standards: Recipes for Reality*. Cambridge MA: MIT Press.

Caplan, J. and J. Torpey (eds). 2001. *Documenting Individual Identity: The Development of State Practices in the Modern World*. Princeton: Princeton University Press.

Clarke, A.E. and J.H. Fujimura (eds). 1992. *The Right Tools for the Job: At Work in Twentieth-Century Life Sciences*. Princeton: Princeton University Press.

Clifford, J. and G.E. Marcus (eds). 1986. *Writing Culture: The Poetics and Politics of Ethnography*. Berkeley: University of California Press.

Cohn, B.S. 1996. *Colonialism and Its Forms of Knowledge: The British in India*. Princeton: Princeton University Press.

Day, R.E. 2001. 'Totality and Representation: a History of Knowledge Management through European Documentation, Critical Modernity, and Post-Fordism', *Journal of the American Society for Information Science and Technology* 53: 725–735.

De Certeau, M. 1984. *The Practice of Everyday Life*, trans. S.F. Rendall. Berkeley: University of California Press.

Dery, D. 1998. '"Papereality" and Learning in Bureaucratic Organizations', *Administration and Society* 29: 677–689.

Flanzreich, G.L. 1993. 'The Library Bureau and Office Technology', *Libraries and Culture* 28: 402–429.

Foucault, M. 1979. *Discipline and Punish: The Birth of the Prison*, trans. A. Sheridan. New York: Vintage.

Gardey, D. 2008. *Écrire, Calculer, Classer: Comment une revolution de papier a transformé les sociétés contemporaines, 1800-1940*. Paris: La Découverte.

Gitelman, L. 2014. *Paper Knowledge: Toward a Media History of Documents*. Durham NC: Duke University Press.

Hacking, I. 1986. 'Making up People' in T.C. Heller (ed.), *Reconstructing Individualism: Autonomy, Individuality and the Self in Western Thought*. Stanford: Stanford University Press, pp. 222–236.

Hawkins, S. 2002. *Writing and Colonialism in Northern Ghana: The Encounter Between the LoDagaa and the 'World on Paper'.* Toronto: University of Toronto Press.

Hull, M.S. 2012a. 'Documents and Bureaucracy', *Annual Review of Anthropology* 41: 251–267.

—— 2012b. *Government of Paper: The Materiality of Bureaucracy in Urban Pakistan.* Berkeley: University of California Press.

Kafka, B. 2012. *The Demon of Writing: Powers and Failures of Paperwork.* New York: Zone Books.

Krajewski, M. 2011. *Paper Machines: About Cards and Catalogs, 1548-1929,* trans. P. Krapp. Cambridge MA: MIT Press.

Latour, B. 1986. 'Visualization and Cognition: Thinking with Eyes and Hands', in H. Kuklick and E. Long (eds), *Knowledge and Society: Studies in the Sociology of Culture Past and Present* 6. Greenwich CT: JAI Press, pp. 1–40.

—— 2009. *The Making of Law: An Ethnography of the Conseil d'État,* trans. M. Brilman and A. Pottage. Cambridge: Polity.

Latour, B. and S. Woolgar. 1979. *Laboratory Life: The Construction of Scientific Facts.* Beverley Hills: Sage.

Lemov, R. 2016. 'Anthropological Data in Danger, c. 1941-1965', in F. Vidal and N. Dias (eds), *Endangerment, Biodiversity and Culture.* London: Routledge, pp. 87–111.

Lipphardt, V. and M. Sommer. 2015. 'Visibility Matters: Diagrammatic Renderings of Human Evolution and Diversity in Physical, Serological and Molecular Anthropology', *History of the Human Sciences* 28: 3–16.

Morris-Reich, A. 2013. 'Anthropology, Standardization and Measurement: Rudolf Martin and Anthropometric Photography', *British Journal of the History of Science* 46: 487–516.

Mueggler, E. 2011. *The Paper Road: Archive and Experience in the Botanical Exploration of West China and Tibet.* Berkeley: University of California Press.

Müller, L. 2014. *White Magic: The Age of Paper,* trans. Jessica Spengler. Cambridge: Polity Press.

Ogburn, M. 2007. *Indian Ink: Script and Print in the Making of the English East India Company.* Chicago: University of Chicago Press.

Riles, A. (ed.). 2006. *Documents: Artifacts of Modern Knowledge.* Ann Arbor: University of Michigan Press.

Robertson, C. 2010. *The Passport in America: The History of a Document.* New York: Oxford University Press.

—— 2015. 'Paper, Information, and Identity in 1920s America', *Information and Culture: A Journal of History* 50: 392–416.

Rotella, E.J. 1981. 'The Transformation of the American Office: Changes in Employment and Technology', *Journal of Economic History* 41: 51–57.

Schaffer, S. 1998. 'The Leviathan of Parsonstown: Literary Technology and Scientific Representation', in T. Lenoir (ed.), *Inscribing Science: Scientific Texts and the Materiality of Communication.* Stanford: Stanford University Press, pp. 182–222.

Shapin, S. 1984. 'Pump and Circumstance: Robert Boyle's Literary Technology', *Social Studies of Science* 14: 481–520.

—— 1989. 'The Invisible Technician', *The American Scientist* 77: 554–563.

Shapin, S. and S. Schaffer. 1984. *Leviathan and the Air-Pump: Hobbes, Boyle, and the Experimental Life.* Princeton: Princeton University Press.

Slaton, A. and J. Abbate. 2001. 'The Hidden Lives of Standards: Technical Prescription and the Transformation of Work in America', in M.T. Allen and G. Hecht (eds), *Technologies of Power.* Cambridge MA: MIT Press, pp. 95–143.

Star, S.L. and M. Lampland. 2009. 'Reckoning with Standards', in M. Lampland and S.L. Star (eds), *Standards and Their Stories: How Quantifying, Classifying, and Formalizing Practices Shape Everyday Life*. Ithaca: Cornell University Press, pp. 3–24.

Starn, O. (ed.). 2015. *Writing Culture and the Life of Anthropology*. Durham NC: Duke University Press.

Stoler, A.L. 2009. *Along the Archival Grain: Epistemic Anxieties and Colonial Common Sense*. Princeton: Princeton University Press.

Thévenot, L. 1984. 'Rules and Implements: Investments in Forms', *Social Science Information* 23: 1–45.

——— 2009. 'Governing Life by Standards: a View from Engagements', *Social Studies of Science* 39: 793–813.

Torpey, J. 2000. *The Invention of the Passport: Surveillance, Citizenship, and the State*. Cambridge: Cambridge University Press.

Vismann, C. 2008. *Files: Law and Media Technology*, trans. G. Winthrop-Young. Stanford: Stanford University Press.

Weber, Max. 1978 [1922]. *Economy and Society: An Outline of Interpretive Sociology*, eds. G. Roth and C. Wittich, trans. E. Fischoff. Berkeley: University of California Press.

Widmer, A. 2008. 'The Effects of Elusive Knowledge: Census, Health Laws and Inconsistently Modern Subjects in Early Colonial Vanuatu', *Journal of Legal Anthropology* 1: 92–116.

Yates, J.A. 1989. *Control Through Communication: The Rise of System in American Management*. Baltimore: Johns Hopkins University Press.

Index